CLIMATE CHANGE
The Hoax of CO_2 Revealed

Dr. Robert E. Marx

CLIMATE CHANGE
The Hoax of CO_2 Revealed

Dr. Robert E. Marx

Copyright © 2024 by Dr. Robert E. Marx

All Rights Reserved

Hollow Man Publishing
Edited by Spring Goodhart

All rights reserved. No part of this publication may be reproduced, distributed, or transmitted in any form or by any means, including photocopying, recording, or other electronic or mechanical methods, without the prior written permission of the author, except in the case of brief quotations embodied in critical reviews and certain other noncommercial uses permitted by copyright law.

ISBN: ISBN: 979-8322-5447-15

Printed in the United States of America.

Acknowledgements

Maria Ruiz: Research Assistant - Executive Assistant. Navigated through the manuscript references and procured images. Indispensable

Jennifer Idol - Underwater Designer: Complex illustrations and graphs portrayed in an understandable and precise manner.

Disclosure

Dr. Marx has no financial interest or connection to any component of the fossil fuel industry or the cattle-dairy industry.

Dedication

This book is dedicated to S. Fred Singer and Michael Shellenberger for leading the way in the real science of climate change. I have personally never met either and they have never met me, but I have the deepest respect and admiration for their contributions to the real science of climate change.

S. Fred Singer published the original science exposing the fraud of Climate Change in his book; Hot Talk, Cold Science now revised and in its third edition. He has been a pioneer in explaining the real story of climate change using proven science facts and vetted data. His passing on April 5, 2020 was a loss for all who respect the truth.

Michael Shellenberger is a true and original environmentalist who Time Magazine named as a "Hero of the Environment." Through his teachings and in his book, "Apocalypse Never", he exposed many of the mistruths of the climate change activists and how those mistruths have come to actually hurt the environment. Today he steadfastly supports true environmental causes and activities with actions not just words, while exposing the misgivings of the Climate Change Obsessed.

Climate Change: The Hoax of CO_2 Revealed

ORIGINAL INTRODUCTION

Climate Change: The Real Story

Our local climate changes throughout each 24-hour day and also changes throughout the year according to the four seasons in most parts of the world. This day-to-day and seasonal change has been going on for billions of years.

However, the climates change of concern to earth's population today is more incremental, gradual, and global. Therefore, to understand climate change and perhaps develop agreed-upon strategies to deal with it, we need to think in four dimensions. That is, the recognized three dimensions of length, width, and depth but now also the fourth dimension; time. Climate change in *Climate Change The Real Story* focuses on global warming and sea level rise as its main constituents but does not ignore its impact on forest ecosystems, extreme weather patterns, human and animal populations, health concerns, etc. as its secondary and tertiary effects.

For political expediency, climate change has often been put into an apocalyptic alarmist no-win framework by some [1,2] and has spawned books such as *Unthinkable Earth Life After Warming*[3] and the *Sixth Extinction: An Unnatural History*[4], as well as a drive for a *Carbon Tax*.[5] Both of these books and several journal articles project what could happen to weather patterns[6,] what could happen to animal species,[7] what could happen to forests,[8] what could happen to coral,[9] what could cause the extinction of humans,[10] The operative word is could, which identifies them as only opinionated projections. This has been correctly pointed out as lacking any real scientific evidence by the International Union for Conservation of Nature (ICUN)[11] and other publications[12] as well as books such as *Apocalypse Never: Why Environmental Alarmism Hurts Us All*.[13] This debate has been expanded with the sprinkling of misinformation, exaggeration, and self-serving projections by the political media complex[10] leaving

Climate Change: The Hoax of CO_2 Revealed

you and me without the factual information to make up our own minds.

The central problem is that we have not completely understood nor brought into the discussion the main driving forces of global warming and sea level rise. Instead, we have "copped" out so to speak to attribute all of it to carbon, particularly atmospheric carbon dioxide (CO_2). I wrote *Climate Change The Real Story* to present a coalescence of known scientific facts to elucidate the many nonhuman made (up to eight) driving forces of our current climate change more significant than CO_2, the pace that they are proceeding and most importantly the degree of their contribution. You may be surprised to learn of the astrophysical contribution to climate change as well as that of volcanoes, undersea heated vents, tectonic plate movements, land rebounds from the last ice age, ocean landfills, meteorites, and river runoff among others. *Climate Change The Real Story* also identifies the many human made driving forces to our climate change including CO_2, human encroachment into our oceans, the decrease in ocean water clarity, human population explosion, pet and farm animal

Dr. Robert Marx

increased numbers, forest fires, and particularly deforestation among others in a manner that others have not brought forward in a cohesive discussion.

The current political media climate of climate change is one of finger pointing and blame. Each side hoping to obtain an economic gain or a gain in political power. However, few positive initiatives have been advanced to mitigate or adapt to the climate change we are experiencing other than to reduce CO_2 emissions which real science indicates will have little impact.[12, 15] Instead, they paint a doomsday scenario that they ask us to accept, ignoring our species' remarkable ability to confront, adapt to, and resolve crises. *Climate Change The Real Story* takes the opposite perspective and points to several worthwhile initiatives. It advances the list of driving forces humankind cannot change, but also advances practical solutions and far thinking methods to indeed mitigate and adapt to an ever-changing world. Some of these solutions are obvious, such as a concerted reforestation effort and cleaning up of the plastic trash from our oceans as well as reduced use of plastic and paper. However, other less obvious

Climate Change: The Hoax of CO$_2$ Revealed

solutions are state, local, and individual reforestation projects, development of an environmental Corps, deepening and expanding inland lakes, re-seeding and regrowth of coral reefs, filtering and reducing river runoff, a return to glass beverage bottles and garbage containers without using plastic bags, levees the likes used in the Netherlands and about New Orleans, and perhaps sun shields in space, among others.

In the first chapter, I ask the reader to put aside their preconceived notions on climate change, as well as their politically biased opinions. The impact of climate change is far above both. In the final chapter, I attempted to rally the reader to take charge of the climate change discussion rather than to leave it to the political media complex. To have faith in the ability we have, as a species demonstrated in the past, to cope with change. After all, the opening words of the United States Constitution are "We the People..." and Abraham Lincoln's famous Gettysburg address includes the phrase that we are a government "of the people, by the people, and for the people". To the readers of *Climate Change The Real*

Dr. Robert Marx

Story, it is up to us, the people, not anyone else. I hope you will find some genuine facts that I have put forth upon which you can base your opinion and base your actions as well as some valuable suggestions on how to proceed in the face of climate change.

INTRODUCTION TO

CLIMATE CHANGE: THE HOAX OF CO_2 REVEALED

A Hoax Akin to the Hoax of the Oil Shortage of the 1970s

As a medical researcher, I was initially curious and accepted the idea of global warming due to CO_2 and greenhouse gases. After all, I was well aware of those certain molecules, particularly H_2O in the form of clouds and humidity as well as CO_2, O_3, and CH_4, which acted as a porous atmospheric blanket to retain heat and literally prevent the planet from freezing. However, after reading Al Gore's and James Hansen's "The Inconvenient Truth" I was dismayed at all the inconsistencies with the known laws of chemistry and physics and the wrong conclusions they derived from their ignorance. Following through, I saw the errors in their computer models and as time went on, every one of their predictions of arctic ice melts, polar bears dying, prolonged droughts etc. did not occur.

This led me to do a systematic research into the physics, chemistry, biology and history of climate change beginning with the earth's known 5 major ice ages and the less severe and shorter ones in - between as well as the warming periods in - between most of which were warmer than that of today.

The outcomes were a disturbing revelation that the rapid apocalyptic climate change advanced by democrat politicians, a get on the bandwagon biased media, corrupt national and international organizations and some paid for scientists was a scam. The scam is to promote the acceptance of a carbon tax, the cap-and-trade concept of carbon credits, frighten the public to vote for certain "save the planet" candidates and for the public to accept the "green new deal" which is actually a very bad deal.

Climate Change: The Hoax of CO$_2$ Revealed

The politically promoted and exaggerated claims of the climate change obsessed should remind us all of the 1970s oil shortage claims that were also promoted by democrat politicians, a "get on the bandwagon" media and some paid for oil experts. History notes that all those who were so sure that the world's oil would run out by 2007 were wrong. Therefore, one should ask where are they now? The outcome of the oil shortage scam was long gas lines, gas rationing, 55 miles per hour speed limits, high unemployment coupled with high inflation termed the misery index and a weakened respect from our allies and adversaries. It was all a ruse based on incomplete data, inaccurate science, and a desire to control the public. Climate change is today's oil shortage ruse.

I have written Climate Change: the Hoax of CO$_2$ Revealed for the average person. It presents the facts, data, and science that the politicians and media have either not been aware of or have purposefully withheld from the public. I have tried to explain such things as astrophysics, solar radiation, the earth's climate history, ice ages, the real greenhouse mechanism and even cow methane producing flatulence in an understandable fashion.

Chapter one exposes the fundamental flaws of the Inconvenient Truth and exposes the fact that a gradual global warming as a rebound from the last ice age and from solar variations is the cause of atmospheric CO_2 rising, not that rising CO_2 is causing global warming.

Chapter two reveals the real sources of a gradual warming planet as solar output variations and the Milankovitch orbital path variations.

Chapter three identifies the many real reasons that our oceans are slowly rising at a rate of one inch per decade, none of which is from fossil fuels or CO_2.

Chapter four identifies how greenhouse gases work and identifies that our planet is not a true closed greenhouse. It is a poor analogy. It also debunks much of the false claims of the climate change obsessed, such as bleached coral, higher tides, and

Climate Change: The Hoax of CO₂ Revealed

drought predictions.

Chapter five exposes fake science and how they alter data and dismiss the current more accurate data from coming to the attention of the public. Regrettably, well-recognized organizations with a legacy of respect, such as NOAA, the EPA, and the UN, are part of the problem.

Chapter six exposes the hypocrisy of the climate change obsessed. That is, if they decry CO_2, they have made no steps to curtail CO_2 other than their vilification of fossil fuels, their mandates for EV cars, and blaming all of us for using natural gas, and gasoline-powered cars, etc. This chapter enlightens them to the many ways they could undertake to control CO_2.

Chapter seven debunks the idea of cow flatulence as a contributor to global warming. Again, it advances known methods to reduce the methane emission from cows by over 80%. It also enlightens them to the fact the 1.4 billion lightning strikes globally burns off most if not all of the highly flammable atmospheric methane, explaining why its concentration is only 0.00017% of the atmosphere. The reader does not need to feel guilty about drinking milk or eating a hamburger or steak.

Chapter eight discusses the benefit of a warmer climate throughout history as well as the increased current productivity of a gradually warming planet.

Chapter nine discusses three real existential threats far more serious than climate change and why. This chapter details why overpopulation, toxic pollution, and more pandemics are the real existential threats.

As you read this book, realize that our climate is warming and sea levels are rising at a very slow pace (1.08 degrees F per century and one inch per decade respectively) but not at all from CO_2 or fossil fuels. As Harvard Professor George Santayana advised us all, "Those who cannot learn from the past are condemned to repeat it". Let us not repeat the oil shortage scam of the 1970s.

Climate Change: The Hoax of CO_2 Revealed

Chapter One

CO_2 is Not the Cause of Global Warming

The message of this book is that climate change, i.e., global warming and sea level rise, are real but CO_2 is a scam. This is not just an opinion, but a fact. One that I will prove to you in this book. This statement is also not just the statement of this author as some sort of a lone wolf or rogue but is also that of the "Global Climate Intelligence Group".[1] A group of 1,609 nongovernment funded scientists that includes Nobel Laureates, Emeritus Professors of Geophysics and Meteorology, as well as distinguished Meteorologists and more.[2] In response to their concerns over the climate misinformation, lack of real science and the exaggerated predictions emanating from agenda driven politicians, a misinformed media and misguided radical environmentalists, they published a fact-based report called the "World Climate Declaration" in 2019.[3] The belief that CO_2 is the driving force of our current very gradual global warming is now exposed as just that, a belief not science, and that the following science should convince all who read it that the United States and much of the world has gone down the wrong path. One that will not affect climate change one bit while seriously affecting the world's economic balance and the everyday lives of many.

There were at least ten "Conferences of the Parties" (COPS's) organized by the United Nations concerning climate change by 2005 and earlier work on earth's temperature rise by Joseph Fourier as early as 1824 and by Savant Arrhenius in 1896.[4] However, it was not until former Vice President Al Gore and James Hansen of NASA made the proclamation that climate change was an "existential threat" to humanity that it became a political ideology.[5] It generated predictions of apocalyptic proportions and a fear of earth's end that has until now overwhelmed real science and mislead a public unaware of the real facts.

Climate Change: The Hoax of CO_2 Revealed

So, what were Al Gore's and James Hansen's errors that resulted in the public's acceptance of a wrong concept?

We can break it down to:

I. **Incorrect and uneducated knowledge concerning physical chemistry**.

Al Gore published his treatise, The "Inconvenient Truth",[5] in 2005 and lectured extensively on it. Perhaps it would have been better to label it the convenient mistruth as Al Gore was later exposed to have heavily invested in oil production, maintained four residences, a yacht and traveled around the globe decrying CO_2 with his CO_2 emitting jet. He also developed a company to trade carbon credits and carbon permits. He was even awarded the Nobel Peace Prize for his unproven and now debunked theory.[6]

In addition to the galling hypocrisy and greed, the science is wrong. In The Inconvenient Truth, he depicted industrial smoke stacks billowing out white

Dr. Robert Marx

and black smoke spiraling into class 5 hurricanes (Fig 1-1). Al Gore apparently did not learn that CO_2 is odorless and colorless. The white part of the smoke he depicted is mostly water vapor arising from that which is being burned and the black part is soot. Soot is a particulate, not a gas, and is known to have the chemical formula (C_3) H.

That is, it is a particulate of fixed carbon which settles as a type of ash and does not contribute to CO_2.

Figure 1-1

Climate Change: The Hoax of CO_2 Revealed

Fig 1-1 *Since CO_2 is odorless and colorless the white and black smoke in Al Gore's 2005 Inconvenient Truth" swirling into class 5 hurricanes is not CO_2. Instead, the white is water vapor from water in whatever was being burned and the black is soot which is a fixed particulate not a gas and not CO_2.*

Certainly, there is some CO_2 emanating from a smoke stack but it is much less than you and I were led to believe.

II. All of Al Gore's and James Hansen's predictions proved to be wrong.

Along with the Inconvenient Truth, Al Gore and James Hansen predicted that all arctic ice would be gone by 2015 causing polar bears to die and their food source to be gone.[7] Actually, as of 2023 there remains plenty of ice in the arctic. It is seven feet thick and the area has not decreased compared to 1983.[7] Also, the census of polar bears and their most common food source, the ringed seal, is greater today than in 2005.

Dr. Robert Marx

After 2005, a flood of other predictions of disastrous consequences related to CO_2 causing global warming came out. All have not come to pass. Examples include:

1. July 2020, The Australian Geographic Publication predicted "the end of snow". However, 2021 and 2022 had above average snow falls.[7]
2. August of 2022. A Bloomberg prediction "the end of snow threatens to upend 76 million Americans lives in the western united states". A few months later, the Sierra Nevada Mountains recorded their second snowiest winter on record.[7]
3. Harvard University Professor of Atmospheric Chemistry. James Anderson "the chance there will be any permanent ice left in the arctic after 2022 is essentially zero.[7] Today, the arctic ice is equal to what it was in the 1980s.[7]
4. Arctic ice has seen an irreversible thinning since 2007 Kerem Yuel, March of 2023.[7] However, a more recent and more direct study by NASA scientist Jay Z Wally stated that the Greenland ice sheets are actually becoming thicker.[7]

Climate Change: The Hoax of CO_2 Revealed

5. My own personal note. Attending a lecture at the Miami Museum of Science, I heard the presenter say that earth will become just like Venus with a mostly CO_2 atmosphere and temperatures at 250 degrees Celsius (482 degrees Fahrenheit). He showed several animated pictures depicting volcanoes and swirling sulfurous acid clouds to illustrate his point. However, when I pointed out to him that humans were never on Venus and that the volcanoes were the source of the sulfurous acid in the CO_2 atmosphere. He had no response. He also apparently did not know that the greatest extinction event in earth's history known as the Permian extinction was caused by repetitive volcanic eruptions over centuries and even millennia 200 million years ago without any of humankind's contribution.

6. June 2008, James Hansen the NASA computer scientist predicted that "In five to ten years the arctic will be free of ice".[7] This mentor of Al Gore was just as wrong in 2008 as Al Gore was in 2005. The arctic ice remains stable today. His computer modeling was wrong. Why? The simple reason is

Dr. Robert Marx

the weakness of all computer-generated models. It is that critical data is often omitted, may be overrepresented or underrepresented, and assumptions (also called guesses) are made along the way all of which affect the final conclusion. Simply stated, garbage in – garbage out. So, what did James Hansen not include or mis-weigh in his modeling?

A. Clouds

B. Ocean currents

C. Volcanoes

D. Undersea vents particularly the mid-Atlantic ridge where a 12,000 miles by three hundred to six hundred miles ridge of active heated vents exists. As a case in point, the week of Nov. 5, 2023 saw an entire new island emerge off the coast of Japan from a similar chain of undersea vents (volcanoes).

E. Solar variations

F. Variations in the orbital path of the earth around the sun

G. He most egregiously used an exaggerated life span for CO_2 in the atmosphere as 100 years.[8]

Climate Change: The Hoax of CO_2 Revealed

This is vastly different from the norm agreed upon by every other scientist as 3.5 years.[9] One can see a computer input of a 96.5 year difference from the normal can skew the result to be what you wanted it to be.

III. Ignorance of the geologic temperature variance of the past 4,000 years.

The historical record of earth's warming and cooling periods is well known through isotopes of oxygen in ice cores and soil cores and is agreed upon by most all climate scientists.[10] Our last of earth's 5 major ice ages ended 12,000 years ago. From that time to the time of Christ, 2,000 years ago, there was a continued gradual warming with minor increases and decreases in temperatures while still warmer than the ice age temperatures. However, there were at least 17 cooling periods ranging from decades to millennia during that 10,000-year period.[18]

Around the time of Christ, there was a noted increased warming trend that lasted from around 200 BCE to 400 AD.[10] This is known as the "Roman

Warm Period". Such a warming period saw average temperatures even above our current temperatures today. This translated into the longer growing seasons, less frost, more rainfall, higher crop yields, all of which supported the prosperity of the Romans, Egyptians, and Greeks at the time. However, this trend reversed itself around 400 AD with a significant cooling period that extended to around 800 AD and is known as the "dark ages cool period".[10] Then a return to another a warm period occurred from 900 AD to 1400 AD termed the "Medieval Warm Period."[10] This period saw the expansion of European art, commerce, and populations once again due to an enhance crop yield from a longer growing season, more rainfall and fewer crop losses due to frost. It also saw the expansion of the Vikings who could now navigate the ice locked fjords and the North Atlantic, which allowed them to develop settlements in Greenland. This is supported by the finding if a 12th Century Viking settlement under the current Greenland glacier indicating the earth's temperature at that latitude was warmer than it is today.[11,12] (Fig 1-2)

Climate Change: The Hoax of CO_2 Revealed

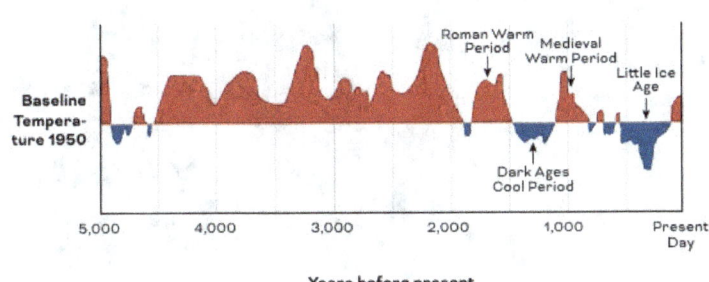

Figure 1-2

Fig 1-2 *Timeline of past global warm and cool periods over the past 5,000 years*

The most important and severe variation in earth's temperature since the last ice age occurred from 1400 AD to 1850 AD. It was a significant cooling trend termed "The Little Ice Age" where the average temperatures were the coldest ever recorded.[13] The 1776 cold winter at Valley Forge with our troops hungry, sick, and cold from one of the most brutal winters ever is no exaggeration (Fig 1-3).

Dr. Robert Marx

Figure 1-3

Fig 1-3 *Valley Forge Winter 1777 - 1778 was noted to be exceptionally cold and was is in the middle of "The Little Ice Age"*

The "Little Ice Age" continued until 1850, when our current warming trend began. However, the current warming trend has not been without its short cool periods between 1850 and the present. There have been at least six, one cooling period occurred between 1975 and 1983 and then again 1999 to 2012.[14] These facts should resonate within us all. These numerous fluctuations in earth's temperatures from biblical times to the present are all unrelated to humans burning fossil fuels and CO_2. The fact that so many dire predictions by even well-educated individuals

Climate Change: The Hoax of CO_2 Revealed

were so wrong should make us suspicious of their own hidden agendas or perhaps they just didn't consider or know of the natural fluctuations both short (decades and centuries long) and long (several centuries to several millennia). In fact, our last ice age peaked 25,000 years ago. At that time, there was a one- and one-half mile glacier extending over all of present-day Canada extending down to the Illinois-Wisconsin border as well as the same latitudes in Europe and Asia.[15] The English Channel was a dry tundra.[15] Today, we find the bones of the terrestrial Pleistocene fauna of the last ice age as well as the tusks of mammoths in depths of 180 to 300 feet. The ocean levels were 300 to 500 feet lower than today. One must ask the question: what did CO_2 and the burning of fossil fuels have to do with that massive glacier melt? Certainly, Fred Flintstone and Barney Rubble were not driving gasoline driven cars (Fig 1-4).

Dr. Robert Marx

Figure 1-4

Fig 1-4 *The Hanna and Barbera prehistoric cartoon characters Fred Flintstone and Barney Rubble use pedal power not fossil fuels to power their "Flintmobile"*

IV. Misunderstanding the Laws of Physics governing gases dissolved in liquids.

Since the early days of climate concerns scientists such as French mathematician Joseph Fourier (1824) have tried to infer a link between atmospheric CO_2 and increased temperatures. This morphed into the term "greenhouse effect" by Swedish chemist Steven Arrhenius in 1896.[4] It is known that earth's temperatures are stabilized by several molecules (H_2O, CO_2, CH_4, O_3, etc.) in a natural pattern akin to a greenhouse to prevent heat loss into space and has

Climate Change: The Hoax of CO_2 Revealed

been responsible for our mild climate since the last ice age. However, it is CO_2 that is now being vilified for a factless apocalyptic global temperature risk that is actually not apocalyptic but gradual. This gradual rise in earth's temperature is known to be caused by other events (the subject of Chapter 2).

What Al Gore, James Hansen and those obsessed with linking earth's current gradual temperature increase to CO_2 didn't realize is the relationship of dissolved gases in a liquid to the temperature of the liquid. After the end of "the Little Ice Age" in 1850, atmospheric temperatures did begin to rise. Al Gore and James Hansen noted that as the industrial revolution took foot hold and the burning of fossil fuels became common with the invention and popularity of automobiles, trucks, and diesel-powered locomotives in the early 1920s. They also noted that atmospheric CO_2 was increasing and wrongly assumed that global warming was due to CO_2. It seemed logical that the increasing atmospheric CO_2 was causing the warming trend. It was accepted as fact and promoted as fact by Al Gore and James

Hansen in the mid 2000s and now by President Joe Biden and Vice President Kamala Harris. None of these individuals looked for a proof of a cause and effect relationship or other factors that were causing the earth to gradually warm. Had they actually studied it, they would have discovered the laws of physics governing gases such as CO_2, O_2, N_2, Ar, O_3, common in our atmosphere. The law of physics states that as the temperature of the liquid rises, it will hold less dissolved gas (Fig 1-5).[16] You can verify this yourself by taking a glass of refrigerated water and leave it out at room temperature or even out in the sun if you prefer. Over time, you will notice bubbles appearing on the side of the glass (Fig 1-6). These bubbles are O_2, CO_2, N_2 etc. dissolved in the water that are now being released. Related to global warming, this has been going on since 1850. The warming trend caused by natural factors the likes of which ended the last major ice age has caused and is continuing to cause the CO_2 dissolved in the greatest reservoir of CO_2 on the planet, the oceans to release CO_2. We may not know which came first, the chicken or the egg, but we do know that global warming

Climate Change: The Hoax of CO_2 Revealed

preceded a rise in atmospheric CO_2 (Fig 1-7). If you look at the curve presented in figure 1-7, you will notice that the rise in temperature and the rise in CO_2 taken from ice core samples as well as direct measurement from 1850 to the present are parallel while the average of CO_2 computer climate models predicts a much exaggerate and rapid temperature rise that has not occurred.[17] This curve indicates that the documented rise in atmospheric CO_2 is real but unrelated to fossil fuels. It is related to the slow rate of ocean warming releasing CO_2. The fact is that CO_2 rise is not the cause of global warming. Global warming is the cause of CO_2 rise. The vastness of the oceans (70% of earth's surface) and its depth dwarfs mankind's emissions of CO_2.

To prove this, one has to only look at established facts. That is, our oceans contain a total of 3.56 x 10^{19} gallons of water. To put this vast number in perspective 10^9 is a billion, 10^{12} is a trillion, 10^{15} is a quadrillion and 10^{18} is a quintillion. Therefore, our oceans contain 10 quintillion gallons of water. The average temperature of our oceans is said to be 70 degrees Fahrenheit. Fig 1-6 indicates that at that temperature our oceans contain 240,000 trillion tons of CO_2. The average temperature of our oceans has increased just 0.1 degrees Fahrenheit due to solar activity and earth's orbital variations in the last 75 years. This has release 31.25 trillion tons of CO_2 into the atmosphere. This amount dwarfs the estimated 25 billion tons of CO_2 from fossil fuels and is likely much more than from volcanoes. Indeed the dissolved CO_2 from our oceans is the real responsible source of increased atmospheric CO_2 nothing else.

Climate Change: The Hoax of CO₂ Revealed

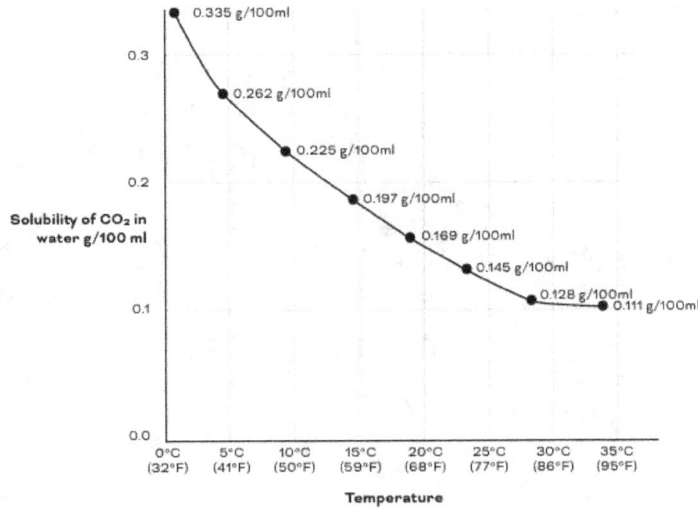

Figure 1-5

Fig 1-5 *The solubility of CO₂ in water rapidly declines as the water temperature rises. Even small increases in the water temperature of the ocean releases a significant amount of CO₂ as illustrated by the steep decline in its solubility curve. That is, an ocean temperature rise of just 0.1 degrees Fahrenheit will release 31.25 Trillion tons of CO₂ into the atmosphere.*

Dr. Robert Marx

Figure 1-6

Fig 1-6 These bubbles represent CO_2, O_2 and N_2 that were dissolved in the cold water at 35 degrees Fahrenheit and came out of solution as the water warmed to room temperature at 72 degrees Fahrenheit. This law explains the atmospheric rise of CO_2 coming from our oceans as our planet's temperature gradually rises and debunks the notion that it the result of fossil fuels.

Climate Change: The Hoax of CO_2 Revealed

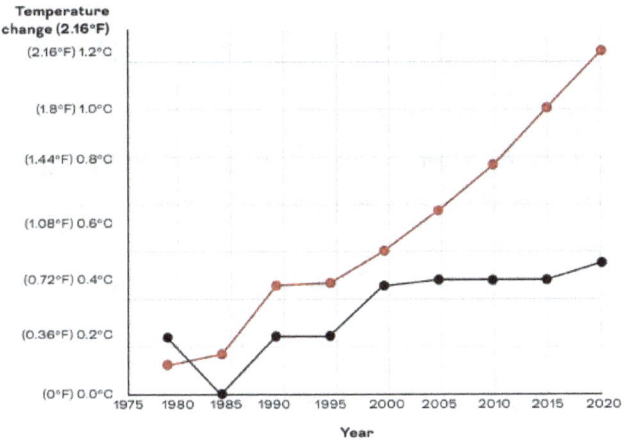

Figure 1.7

- Computer climate models
- Average of satellite and balloon measurements

Fig 1-7 *the computer models predicting run away temperatures have been consistently wrong compared to actual temperature readings leading to a false fear advanced by the climate changed obsessed.*

Dr. Robert Marx

References
Chapter 1

1. https://clintel.org> World Climate Declaration, Jan. 12, 2024
2. https://insideclimatenews.org>News, Aug. 23, 2022
3. https://www.azemews.az>region, Nov. 19, 2023
4. Blair TA, Climatology: General and Regional. Prentice-Hall, 1942, 118
5. Gore A. An Inconvenient Truth: Rodale Press Inc. 2006
6. https://skepticalscience.com>al-gore-inconvenient-truth
7. https://www.theepochtimes.com>epochto>inconvenienttruth. Inconvenient truth:32 predictions proven false. Mar 2023
8. Hansen J, Sato M. 2012. "Paleoclimate Implications for Human Made Climate Change". In Climate Change: Inferences form Paleoclimate and Regional Aspects, edited by A. Berger, F Mesinger, and D. Sijacki, 21-48. Vienna: Springer.

https://doi.org/10.1007/978-3-7091-0973-1_2.

9. IPCC AR-5. 2013, 2014 Climate4 Change 2013 (working group I contribution) and Climate Change 2014 (working groups II and III contributions). Fifth Assessment Report of the Intergovernmental Panel on climate change, Cambridge. Cambridge University Press.

10. Alley RB. The younger dryas cold interval as viewed from Central Greenland. Quaternary Science Reviews 2000,16:213-226.

11. Zwally HJ. Growth of Greenland Ice Sheet: Interpretation Science 1989, Vol. 246, 4937, 1589-1591

12. Singer SF, Legates Dr, Lupo AR. Hot Talk Cold Science Independent Institute. 2021. pp 88-135

13. Singer SF, Avery DT. Unstoppable Global Warming every 1,500 years. Updated and expanded el. Lanham MD: Rowman and Littlefield Publishers Inc. 2008

14. Spencer R. AH Global temperature Update for October. Dr. Roy Spencer (blog). Nov. 2, 2018. Https://www.drroyspencer.com/2018/11/uah-

global temperature-update-for-october-2018-0-22-deg-c/.

15. Kim Ann Zimmerman-Live Contributor Aug. 29, 2017. Pleistocene Epoch: Facts about the last ice age.
https://www.livescience.com/ 40311-pleistocene-epoch.html

16. www.EngineeringToolBox.com. Solubility of gases in water vs temperature.

17. Christy JR. Testimony before the US House Committee on Science, Space and Technology. Mar. 29, 2017

Climate Change: The Hoax of CO_2 Revealed

Chapter 2

The Actual Causes of Global Warming

It is one thing to point out the flaws and misunderstandings of those who are obsessed with CO_2 and fossil fuel burning as the cause of global warming but it is another thing to prove the real cause. However, proving the real cause is actually straightforward. It is the variations in the light and electromagnetic waves from the sun and the variations in the orbital path of the earth around the sun.[1,2]

I. Variation in Solar Output:

Our sun, like planet earth, is 4.57 billion years old and is halfway to its life expectancy of 10 billion years. Since the sun is the root source of all of earth's energy except that from volcanoes, earth's life expectancy is only 2 to 5 days longer than that of the sun. If the sun burns out, the earth will experience a rapid cooling that will encase the earth in ice in less than a week. Therefore, it is good to recall that the earth's temperature is almost totally dependent on the sun, particularly its luminescent output.

The sun is actually one of many stars. Like all stars, it pulsates to a degree with periods of greater and lesser output. The most documented output cycle for the sun is an eleven-year cycle which has been attributed to the El Niño increased temperatures in the Pacific Ocean and western states.[3] Note, a first ever hurricane making landfall in California in 2023 was attributed to the 2023 El Niño, as was the heat wave in 1998 and that of 2023 registered as the hottest July on record. This cycle also creates a La Niña affect which pushes the jet stream north ward over the eastern pacific producing La Niña winters of colder and wetter conditions over Canada and the North and Midwest United States.[4] The La Niña effect is produced by the heat depleted post La Niño Pacific Ocean waters, producing cold air over the northern and Midwestern states. This affect was the cause of the subzero temperatures that affected the

Climate Change: The Hoax of CO₂ Revealed

Republican Iowa primary of 2024 and caused the postponement of an NFL wild card game. However, the sun has several longer and several shorter periods of increased or decreased solar output that are more random.

"The Little Ice Age" between 1400 and 1850 is an example of a decreased solar output. The solar output at that time was 1360.1 watts/m^2, one of the lowest on record.[5] At the end of the little ice age in 1850, the recordings quickly increased to 1361.8 watts/m^2 today. Now a square meter is a small surface area. US citizens would be well to know that a meter is 1.09 yards or 39.7 inches. To gauge this square surface area, one could add 3.7 inches of tape to each of four common yardsticks and place them in a square. You will see that this one-meter square is miniscule compared to the square surface area of the earth. While the difference between "The Little Ice Age" and the current warming period is only 1.7 watts/m^2, the earth's surface area is 510.1 trillion m^2. This amounts to 867 trillion more watts of solar energy being absorbed by the earth today which of course radiates heat back into the atmosphere. This gradual heating has been going on and building up since 1850 as discussed in Chapter one. This gradual heating causes more CO_2 to be released from our oceans than before 1850 as shown in Figure 1-6.

We know that any number of watts hitting the earth's surface will emit and generate heat into the atmosphere. Just try walking barefoot on an asphalt surface or placing your hand on the hood of a car during a mid-day summer afternoon. This is where human beings have increased the warming of the atmosphere. Today, only 3.8 trillion trees are said to remain today giving off shade as compared to 10 trillion trees in 1500.[6] Additionally, humans have replaced green areas with pavement, metal structures and cement all of which radiate more heat into the atmosphere than if it were a field or forest. These man-made structures also retain heat so that they continue radiating their absorbed heat energy long after dusk.

Climate Change: The Hoax of CO_2 Revealed

II. The Shape and Variations of the Earth's Orbital Path Around the Sun:

Contrary to the over simplified drawings of a circular orbital path of the nine planets (if you include Pluto) orbiting the sun, all planets, asteroids and comets orbit the sun in an elliptical path. The orbital path of the earth about the sun is elliptical (also termed eccentricity). However, the path itself has a variation in the form of a tilt (also termed axial tilt) from 0 to 1.5 degrees toward or away from the sun. A third variation of the orbital path is a wobble (also termed precession) in which the orbital path fluctuates and wobbles by the influence of solar winds and the gravitational effects of passing asteroids as well as nearby planets.

The orbital path characteristics exert their influence over 10,000 and 100,000 or millions of years. These orbital path characteristics are called the Milankovitch cycles and are unarguable among all scientists today.[7] These cycles were brought to the attention of the scientific community in 1920 by Serbian geophysicist Milutin Milankovitch who consolidated and confirmed the earlier findings of French Mathematician Joseph Adhemar and Scottish scientist James Croll.[2]

The Milankovitch cycles explain all of our past ice ages and can even explain interglacial warming periods as well as shorter ones such as the Roman and Medieval warming periods and short (centuries long) cooling periods such as "The Little Ice Age".

To understand the strong impact on earth's temperature from the Milankovitch cycles and solar radiation we need to start with the basics.

The Sun and The Earth A01:
The earth makes one complete revolution about its axis in a 24-hour period (called one day). At any given time, one side of the Earth is facing the sun (daytime) and the other half is in the shade (nighttime). The sun's rays make daytime significantly warmer than the night, with fluctuations small (3° to 7° F) or large (30° to 50° F). The swing depends on the time of year and the particular location and surroundings. We can

Climate Change: The Hoax of CO$_2$ Revealed

feel the air warming after sunrise and we begin to feel it cooling just before sunset and after. However, midday (high noon) when the sun is directly overhead, is not the hottest time of the day.

Surfaces contacted by the sun (e.g., our fields, buildings, streets, pavement, metals, etc.) absorb photons from sunlight and radiate heat into the surrounding air. Depending on the heat capacity (the amount of heat it takes to increase the temperature of each material), this delays the temperature rise between one and four hours, so that the hottest time of the day is usually around 2:00 pm to 4:00 pm. Conversely, the coldest time of the night is not midway between sundown and sunrise, but around 2:00 am to 4:00 am due to the heat leaving materials that have been warmed by the sun during the day.

This illustrates solar radiation's (sunlight) powerful influence on earth's temperatures. Specifically, the enormous temperatures within the sun do not travel the average 93 million miles to earth. Consider an airliner traveling at altitude. The temperature data displayed while flying at an altitude of 35,000 feet

Dr. Robert Marx

(6.63 miles) will be -50° F to -65° F. Instead, the light energy of the photons that hit the earth's surface is transferred, in part, to the molecules in the material it strikes. These molecules absorb the energy and vibrate more vigorously, bumping into each other as further light energy is transferred to them resulting in friction and heat radiating into the surrounding air.

In addition to one complete revolution about its axis each day, the earth makes one complete trip around the sun during a one-year period. During this time, the people in the Northern Hemisphere and Southern Hemisphere experience the four seasons. At the equator, seasonal changes do not exist, while areas close to the equator (the tropics) experience only very slight seasonal changes. The tilt of the earth as it spins on its axis accounts for the difference in seasonal changes. This tilt is 23.439 degrees from the equator in each direction at its greatest and marks the Tropic of Cancer in the Northern Hemisphere and the Tropic of Capricorn in the Southern Hemisphere.[9] If you live in the Northern Hemisphere, the maximum tilt toward the sun occurs on June 21st known as the

Climate Change: The Hoax of CO$_2$ Revealed

summer solstice. This is the longest day of the year when the sun shines most directly on the Northern Hemisphere. However, it is not usually the hottest day of the year. The hottest days are found in mid-July to mid-August. This is due once again to the accumulation of heat (heat capacity) in the various manufactured and natural materials on the earth's surface. Water, including that found in lakes and oceans, has a very high heat capacity and gives off the heat slowly. That is, it takes time for the sun's light energy to heat up the water and other elements and time to release it. Therefore, the hottest temperatures are delayed by weeks or months.

After June 21st in the Northern Hemisphere, the earth begins to tilt in the opposite direction. As it does, the Northern Hemisphere is positioned slightly further away from the sun so that the photons in the sun's rays hit the surface less directly and with slightly less energy each day. The resultant effect is one of cooler days which progress to colder days as the tilt moves toward its maximum distance from the sun in the Northern Hemisphere and its closest distance to the sun in the Southern Hemisphere on December 21st. Once again, December 21st may be the shortest day in the Northern Hemisphere but it is not usually the coldest. The coldest days are more often experienced in January or even February. This is also due to the radiation of heat accumulated during the summer months being slowly dissipated and moderating the cold early winter days slightly until the reduced sunlight of winter has dissipated the residual heat maximally.

This review of every day and every year cycles illustrates the enormous influence on the degree and control sunlight has on earth's temperature. It underscores that just a small change in either distance from the sun, the angle of the earth to the sun and the number of man-made materials absorbing sunlight can have on earth's climate.

The daily and yearly earth-sun cycles are not part of the Milankovitch cycles. The Milankovitch cycles takes over from here, identifying that the shape of earth's orbit about the sun is elliptical rather than

Climate Change: The Hoax of CO₂ Revealed

circular. That shape has continued over the millennia and is what Milankovitch termed eccentric.

Eccentricity:

The first part of the Milankovitch cycles is "eccentricity" identifying the elliptical path of the earth around the sun. It is but one of its three components. The other two are orbital axial tilt and orbital wobble. (Fig 2-1) The very elliptical nature of this path explains that the often repeated phrase that the sun is 93 million miles from the earth is wrong. Science recognizes that the closest the earth gets to the sun is 91,419,000 miles and is called parhelion. This occurs on January 3rd each year. The farthest the earth is from the sun is 94,581 million miles and is called aphelion. Aphelion occurs on July 4th each year. This is a 01.7% change in distance each way for an overall 3.4% change each year. If you live in the northern hemisphere, you may be quizzical about January 3rd being a time of the year when you are now closest to the sun. Yet it is the dead of winter and cold. The answer is the 23.439-degree tilt of the earth in which the sun's rays strike the earth at an

Dr. Robert Marx

angle that delivers much less direct energy. We can also at least see that in both hemispheres. This moderates to a degree the cold of each hemisphere's winter and the heat of each summer.

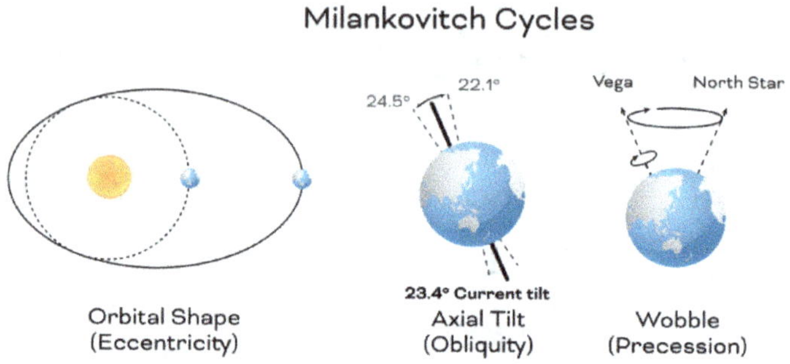

Milankovitch Cycles

Orbital Shape (Eccentricity)   Axial Tilt (Obliquity)   Wobble (Precession)

Figure 2-1

Fig 2-1 *The three components of the Milankovitch cycles: orbital shape (eccentricity), axial tilt (obliquity) and wobble (precession).*

Directly related to the eccentricity component of the Milankovitch cycle is simple mathematics as illustrated in Figure 2-2A. A circular orbital path would be 0.0. Yet it never occurs. The minimal elliptical orbital path of the earth around the sun is actually 0.0034 from a full circle.[8] Although 0.0034 seems like a small number it equates to 316,200 more miles from the sun than the full circle assumed by those obsessed with CO_2. However, even more

Climate Change: The Hoax of CO_2 Revealed

important is the maximum elliptical shape of earth's orbital path around the sun which is 0.058 which amounts to 5,400,000 miles from a full circle and is 17.1 times greater than the minimal elliptical orbital path. Compared to the minimal elliptical distance, the maximal elliptical distance is 5.08 million more miles. Moreover, the time it takes for the earth to travel the extra 5.08 million miles to reach its maximum elliptical distance and 5.08 million miles back is significant and all that time the earth's orbit is further from the sun. Actually, NASA (National Aeronautics and Space Administration and NOAA National Ocean and Atmospheric Administration) misrepresents the difference between these two elliptical paths.[8] Their websites show a bar indicating that the maximal elliptical path is only 3 times larger than the minimal path. Yet simple mathematics (0.058 divided by 0.0034 = 17.058) tells us it is actually 17.1 times larger. This is but one example of how governmental organizations cover up the impact of non-CO_2 contributions to global warming. There are also variations in the eccentricity pathway between the minimal non-circular and the maximal non-circular path (Fig 2-2B). These variations explain the difference in the degree and how much of the earth's surface is covered by glaciers during an ice age. An obvious example of this is the difference between our most recent ice age which covered 30% of the Earth's surface and that of "Snowball Earth", 850 to 635 million years ago, which covered more than 90% of the earth's surface.[9]

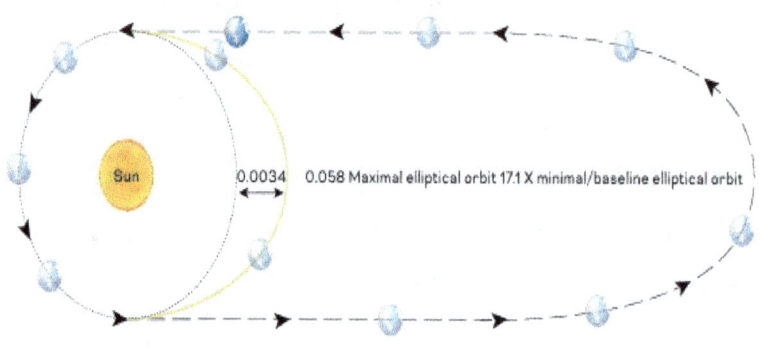

Figure 2-2A

························ Theoretical full circle
─────────── Minimal elliptical orbit
── ── ── Maximal elliptical orbit

Fig 2-2A *The elliptical path of earth's orbit around the sun varies from a true circle. Its minimum variation is 0.0034 from a true circle. Its maximum variation is 0.058 from a true circle, a 17.1 times difference.*

Climate Change: The Hoax of CO_2 Revealed

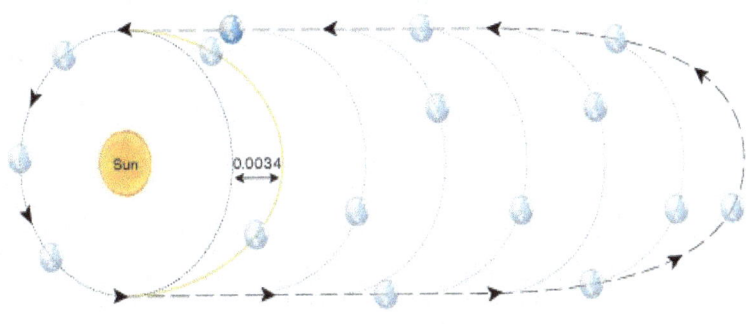

Figure 2-2B

.... Theoretical full circle
———— Minimal elliptical orbit
— — — Maximal elliptical orbit

Fig 2-2B *Variations between the minimal and maximal eccentricity from a true circular path explains the difference in the length, degree and extension over the earth's surface between earth's five major ice ages.*

Understanding just this component of the Milankovitch cycles explains earth's known five major ice ages.

The Hanoian: 2.4 to 2.1 billion years ago

The Cryogenian: 850 to 635 million years ago

The Andean-Saharan: 460-430 million years ago

The Karoo: 360-260 million years ago

The Quaternary: 2.6 million years ago

It would be noteworthy to point out that climate scientists state that the earth is still in an ice age despite the mild temperatures we enjoy today and the claims of climate alarmists that the earth is boiling. The fact is that we are indeed in a post ice age or interglacial period. The last ice age began warming 25,000 years ago and ended 12,000 years ago allowing we homo sapiens to colonize the entire earth. The natural ending of the last ice age was due to the Milankovitch cycling of the earth's orbital path back to the 0.0034 elliptical baseline without a boost from fossil fuel burning and CO_2. This may actually be the root cause of our global warming. The Milankovitch cycle's three components (eccentricity, axial tilt and wobble) were the main driving forces that created our last ice age as it did the other four ice ages, as well as eventually melting each of them. The formation and melting of each ice age were all done without fossil fuel burning. In fact, the Milankovitch cycles can even explain the most severe ice age of all. The Cryogenian ice age that has been termed "snowball earth". This ice age was the most severe and widespread. In the snowball earth ice age, 90% of the earth was covered by glaciers as compared to our last ice age when only 30% of the earth was covered by

Climate Change: The Hoax of CO_2 Revealed

glaciers. It was a time when the orbital path of the earth stayed close to or at the 0.058 maximum ellipse from the elliptical minimum of 0.0034 for millions of years.

The mere fact that the orbital path of the earth changes back and forth over 100,000 or more years explains not only all of our ice ages and interglacial warming periods but also the magnitude and length of its effect regardless of humans (Fig 2-3, Fig 2-4, Fig 2-5)

(Earth's orbital path enters out of a prolonged ice age into a warming trend)

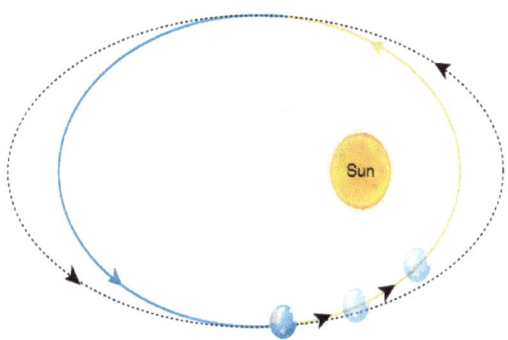

Figure 2-3

Fig 2-3 *Earth's positions in its elliptical path as it exits from an ice age and then proceeds toward a warming period. This represents most closely the position of the earth today as earth is exiting from the last ice age.*

Dr. Robert Marx

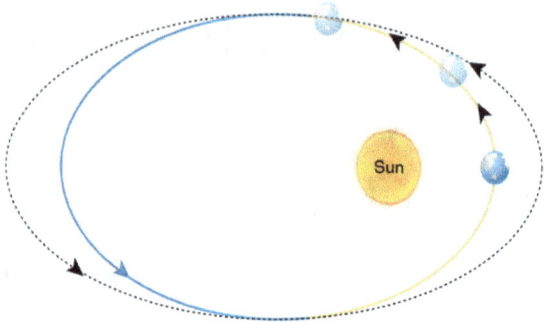

Figure 2-4

***Fig* 2-4** *Earth's position during its warmest long-term period as it occurred during the Jurassic and Cretaceous periods of the dinosaurs.*

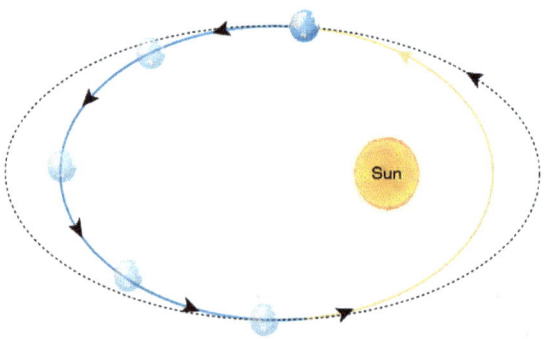

Figure 2-5

***Fig* 2-5** *Earth's position and path as it enters and proceeds through a prolonged ice age.*

Climate Change: The Hoax of CO_2 Revealed

Axial Tilt (obliquity)

The second component of the Milankovitch cycles is axial tilt. It is not the tilt of the earth directly but the tilt of the orbital path, which in turn tilts the earth either toward the sun or away from the sun. This tilting is totally unrelated to the tilting of the earth itself which creates our seasons. The tilt of the orbital path is 1.5 degrees either toward or away from the sun but may last decades, centuries or millennia. It is a three-degree swing. One can conceptualize this tilt as similar to the bank of a race track such as the Indianapolis racetrack which has a 9-degree bank or that of the Daytona speedway, which has a 21-degree bank. If the tilt is toward the sun, we can picture ourselves viewing the race as the sun in the infield. We will see drivers and the top of the race car more directly as it goes by. If the tilt is 0 degrees (neutral) we will see the driver and the driver's side of the car directly but only a small portion of the top of the car. If the tilt is 1.5 degrees away from the sun at the infield, we will see less of the driver as well as less of the driver's side of the car and none of the top of the car.

This axial tilt can explain the centuries long warming periods during the middle of an ice age and the decades or centuries long cooling periods during the periods of an interglacial warming period. The chain of cooling periods followed by warming periods as previously noted spanning several centuries i.e., cooling period 600 BCE to 200 BCE, Roman warming period 200 BCE to 400 AD, the dark ages cooling period 400 AD to 900 AD, medieval warming period 900 AD to 1400 AD. "The Little Ice Age" cooling period 1400 AD to 1850 AD and our current warming period 1850 to the present are examples of the Milankovitch cycles with an axial tilt and wobble unrelated to fossil fuels and CO_2.

Wobble (Wobble)

The earth's orbital path around the sun, like the sun itself, and the other two components of the Milankovitch cycles, is not outside the influence of external forces and is therefore not rigid. Stellar winds, the gravitational effects of passing asteroids and the gravitational forces of nearby planets as we pass closer to them create a wobble in the path. This can result in an enhancement of a warming trend or a dampening of it as well as that of a cooling trend. Together with the short term increases and decreases in the very output of the sun can explain the decades long increases or decreases in earth's temperature as occurred between 1975 and 1983 and 2005 and 2012.

It is my hope that I have distilled down the complexities of the Milankovitch cycles to an understandable form. After reading this chapter, you may feel as the brilliant nuclear physicist Enrico Fermi felt after attending a colleague's lecture on an intricate and complex subject. "At first, I was confused on the subject. I am now still confused but at a higher level"!

Dr. Robert Marx

References
Chapter 2

1. Laskar J, Fienga A, Gastineau M, Manche H (2011) La 2010: A New Orbital Solution for the Long Term. Motion of the Earth. Astronomy & Astrophysics. 532 (A889):A89.arxiv:1103 1084. Bibcode:2011
A&A... 532A.89Ldpo.10.1051/0004-
6361/201 116836.S2CTD10990456
2. Marx RE. Climate Change: The Real Story. Dorrance Publishing 2022 pp19-25
3. https:/ /www.nationalgeographic.org>elNiño
4. https://climate.nasa.gov>explore>ask-nasa-climate
5. https:/ /www.pnel.noaa.gov>elNiño>whatislaNiña
6. Steve Connor: Independent. Sep. 2, 2005. Earth has lost more than half its trees since humans began cutting them down. Independent.co
7. https:/

Climate Change: The Hoax of CO₂ Revealed

/earthhow.com>atmosphere>climatechange. What are the three Milankovitch cycles?

8. https:// climate.nasa.gov>news>milankovitich-orbital-... Milankovitch (orbital) Cycles and Their Role in Earth's Climate, News, Feb. 27, 2020

9. Riell R, et al :Climate Cycles During a Neoproterozoic "Snowball" Glacial Epoch". Geology 2007, 35(4), 299-302. Biocode:2007 Geo... 35 ... 299R doi:10. 1 130/6234000A.

Dr. Robert Marx

Chapter 3

The Rising Seas
Ice Age Rebound

The over simplistic mind set of the CO_2 obsessed is that fossil fuels are burned – CO_2 is given off - temperatures rise - glaciers and ice packs melt - sea levels rise. It is much more complex than that.

It is difficult to confirm sea level rise because of ocean currents, tidal changes, winds, and time of year. There is also no reliable baseline to compare. There is a one inch per decade sea level rise agreed upon by most scientists today. The key question is "What is the cause of this sea level rise?" The answer is not one but a multiplicity of contributions all unrelated to CO_2.

We need to begin with earth's last ice age, which created glaciers covering both poles and beyond. The ice sheet over North America was 1.5 miles thick and extended over all of today's Canada and the Northern

Climate Change: The Hoax of CO_2 Revealed

United States to the Illinois-Wisconsin border as well as the equal latitudes in Asia and Europe[1] (Fig 3-1)

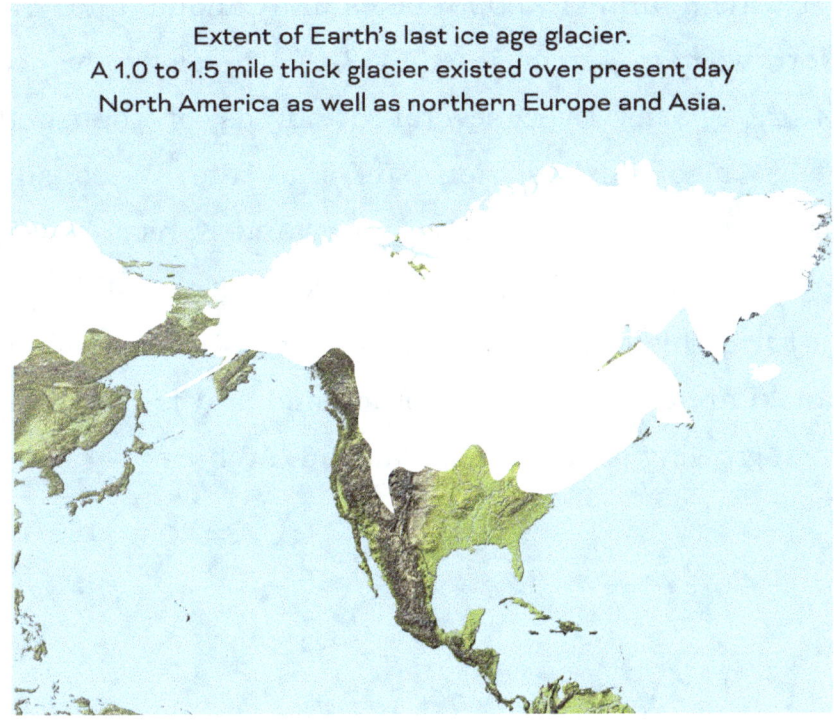

Figure 3-1

Fig 3-1 *During the last ice age 30% of the earth was covered in a glacier. In North America the glacier extended as far south as the Illinois-Wisconsin border.*

The weight of such an ice sheet compressed the ground beneath it and in certain spots gouged out huge depressions that when it melted created our five great lakes, Hudson Bay and five other massive

Dr. Robert Marx

Canadian lakes nearly the size of the great lakes. It also created the 10,000 lakes of Minnesota and Wisconsin and 100,000 lakes in Canada and the Northwestern Territories[2] (Fig 3-2). The slow advance of the glacier over several hundreds of thousand years also pushed rich soil into the Wisconsin, Illinois, Indiana, Nebraska, Iowa, and Kansas areas, creating the bread basket crop yields of the Midwest. In turn, this created what is known as the Canadian shield area where the soil originated and is now base rock growing jack pine, poplar and birch trees.

Climate Change: The Hoax of CO_2 Revealed

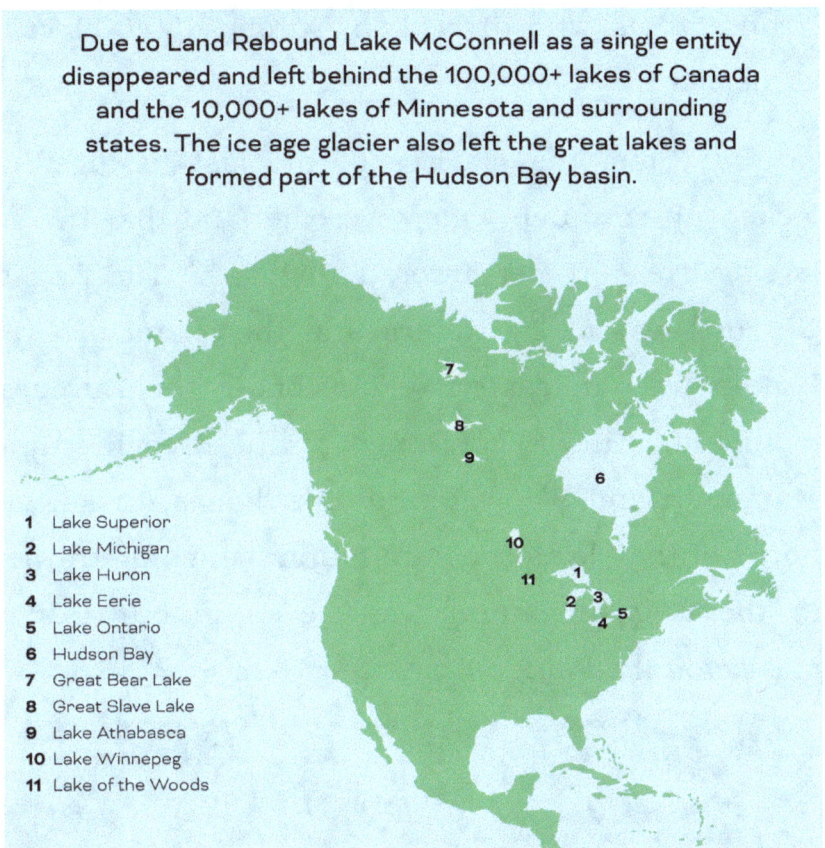

Figure 3-2

Fig 3-2 The eleven "post glacier lakes" in North America were created by the melt that occurred at the end of earth's last ice age.

During this last ice age, homo sapiens emerged out of Africa about 80,000 to 120,000 years ago to set up tribal units in central Asia and the Middle East.[3] They first arrived in Europe a mere 45,000 to 54,000 years ago and were confronted by the already established

Dr. Robert Marx

Neanderthals who arrived in Europe 400,000 years earlier.[4]

At that time, the glaciers of the last ice age had locked up so much water over the land that the sea levels were 400 feet lower than today.[5] The English Channel was a dry tundra and the island of Great Britain was an eastern extension of the European mainland.[6] This is confirmed by the dredging up of Mammoth and Wooly Rhinoceros' bones, tusks, and horns of the Pleistocene era fauna by fishing trawlers in the English Channel and even in our Gulf of Mexico off the Florida Coast[7] Fig 3-3).

Figure 3-3

Climate Change: The Hoax of CO_2 Revealed

Fig 3-3 *A mammoth skull and tusks dredged up in the North Sea between England and the Netherlands identifies that during the last ice age this part of the North Sea and the English Channel were a dry tundra. This area is now 200 to 400 feet deep*

Approximately 25,000 years ago, the last ice age began to wane due to the Milankovitch cycles and the formation of the Isthmus of Panama that occurred 40,000 to 60,000 years ago.

Although scientists state that we are still in an ice age, the real ice age is said to have ended 12,000 years ago when the sea levels and the remaining glaciers were about as they are today. The 13,000-year period that ended the last ice age saw the planet gradually warming unrelated to any fossil fuel burning and CO_2. The slow melting of the last glacier and most of the Arctic Ocean's ice pack was accomplished during those 13,000 years. The melt was accelerated by the formation of the Isthmus of Panama, which closed off the exchange of warm water of the Caribbean basin with the warm waters of the Pacific Ocean at a southern latitude. The creation of the Isthmus of Panama was caused by volcanic activity and tectonic plate movement.[8] In doing so, it created what we now refer to as the Gulfstream in the Atlantic and the Humboldt current in the Pacific. The Gulfstream carried, for the first time in earth's known history, 85-to-90-degree Fahrenheit water northward to end up in the arctic ice pack and Greenland area. It continues to do this to this very day. It is why, today, we have tropical reefs and tropical fish around Bermuda when it has nearly the same latitude as Charleston, South Carolina (32.32 North). (Fig 3-4 and Fig 3-5)[9]

Climate Change: The Hoax of CO₂ Revealed

Figure 3-4

Fig 3-4 *During the last ice age the complete Isthmus of Panama was not yet formed. Warm water flowed freely between the Caribbean Sea and the Southern Atlantic Ocean and the Pacific Ocean diverting it away from the northern ice pack, Iceland and Greenland.*

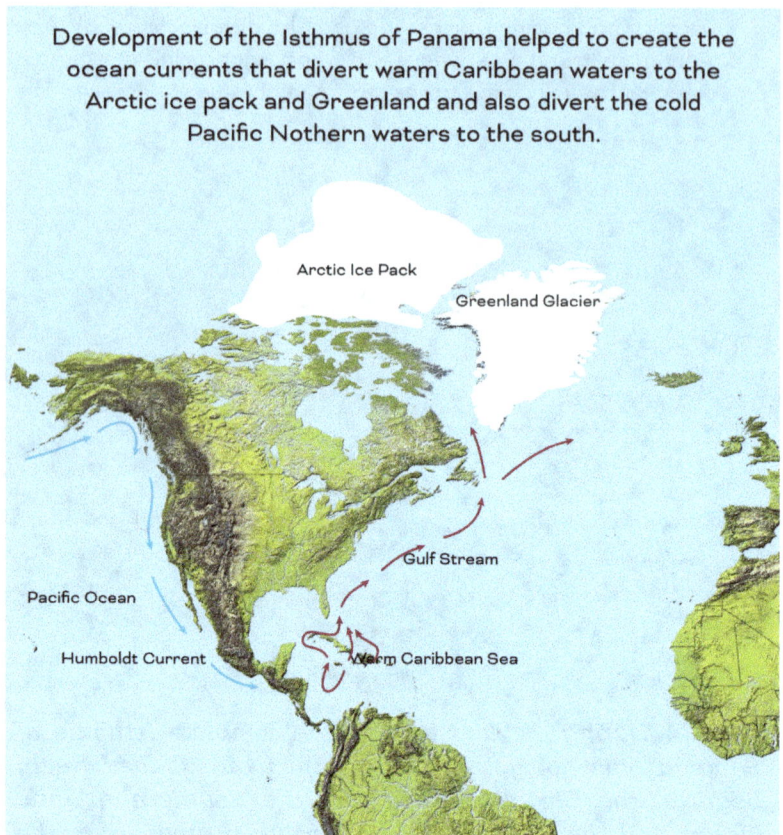

Figure 3-5

Fig 3-5 *Once the Isthmus of Panama completely formed it created the Atlantic gulf stream current that carries warm Caribbean basin waters to the Northern ice pack, Iceland and Greenland.*

This event also created the Humboldt current in the Pacific which transfers cold Alaskan waters southward along our western coast.[10] This is why tropical cyclones (hurricanes) off the coast of

Climate Change: The Hoax of CO_2 Revealed

California are extremely rare while the East Coast of the United States has tropical storms and fully developed hurricanes each year. It is also why we don't see tropical reefs off the California Coast or Baja California. The water is too cold. However, we do see cold water requiring Kelp forests there instead of coral.

When the ice age glacier melted 12,000 years ago, sea levels rose but not to where they are today. Much of the glacier melt stayed in the depressed areas of North America, Europe, and Asia from the very weight of the glacier. In North America, 8,000 years ago, all of Canada east of the Rocky Mountains (Saskatchewan, Alberta, Manitoba, Ontario and even some of Quebec) was a huge inland lake. Geologist have dubbed it "Lake McConell (Fig 3-6).[11]

Dr. Robert Marx

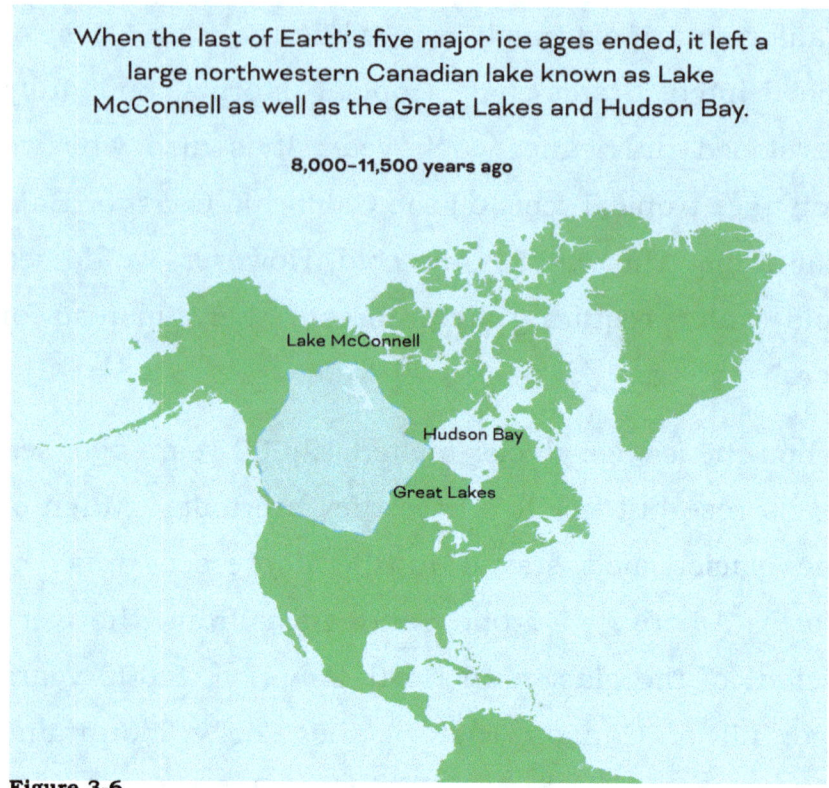

Figure 3-6

Fig 3-6 *The melting of the last ice age left a vast freshwater basin in Canada east of the Rockies known as "Lake McConnell". It has since drained into the oceans leaving over 100,000 post glacier lakes behind including our five Great Lakes.*

In fact, the Cree and Ojibwa natives of mid Canada talk of this lake in their lore handed down through the generations.[12] This lake drained into our oceans over the next 7,000 years via the Columbia and Yukon River systems in the west and the Great

Climate Change: The Hoax of CO₂ Revealed

Lakes, Saint Lawrence, and Mississippi river systems in the east. This brought the sea levels to just about where they are today. The proof of this is illustrated in Fig 3-7 showing the lake level on the rocks of a Manitoba Lake 8,000 years ago compared to its current level today.

Figure 3-7

Fig 3-7 *The water mark of Lake McConnell can be seen 3 to 4 feet above the average water mark in this and most every other Canadian lake.*

Although one may argue that this photo was taken after a particularly low Canadian winter snowfall and limited rain, it was not. I personally took this and other such photographs every year for the past 15 years through both low and highwater seasons. No highwater season ever reached the level mark of Lake McConnel 8,000 years ago. The three-to-four-foot decrease in these 100,000 plus lake levels over the past 8,000 years is an enormous amount of water and is continuing today. What caused the initial formation of Lake McConell was the ice age glacier melt. What caused this massive lake to drain into the oceans to further raise the sea level is the rebound from the weight of the glacier itself. One-and one-half miles of ice packed over 200,000 or more years has an incalculable weight. It compressed the ground enough to form potholes called post glacial lakes, but it also compressed the bedrock. Over the next 7,000 years, the land slowly rebounded, raising the bedrock to initiate a flow that drained these lakes.[13] The aforementioned 100,000 lakes in Canada and the 10,000 lakes in Minnesota and Wisconsin as well as our five great lakes are left over remnants of Lake McConnel.[14] This rebound effect is still going on today as more water is being drained out through the Yukon and Columbia River systems and the

Climate Change: The Hoax of CO_2 Revealed

Mississippi and Saint Lawrence River systems than is being refilled by snow and rain. Evidence of this can be seen in 1,000- to 1,500-year-old petroglyphs in the lakes of Ontario Canada.[15] I have personally viewed several in Eagle Lake Ontario, Canada. Although somewhat faded from years of sun, snow, and ice, they are seen on flat vertical rock cliffs, 7 feet above the current water line. One is hard pressed to understand how these Native Americans, known for their medium to short stature could stand up in canoes and reach way above their heads to paint such intricate details of birds, canoes, wildlife, etc. unless the water level was significantly higher than it is today. Further support of a continued land elevation as a rebound from the last ice age is that the shores of Hudson Bay have receded 100 feet in the past 100 years. This enormous amount of water flowed directly into the Arctic and North Atlantic oceans as the basin of Hudson Bay slowly rose.

The fact of the matter is that the drainage of the lakes left over from lake McConnell continues to this day from the continuing rebound of the lands compressed by the last glacier. If we include the same effect of the last ice age occurring in Europe and Asia, the Black Sea, Caspian Sea, the large lakes in northwestern Russia and Kazakhstan area are also adding to sea level rise.

Archimedes Principle:

Most of us are still in awe when we see a super tanker weighing over 400,000 tons floating, not sinking despite its enormous weight. It floats due to Archimedes' principle, which simply states that "a body in a fluid will gain an upward thrust (float) equal to the weight of the fluid displaced".[16] This principle needs to be recalled as we move ahead in this chapter and realize the importance of the amount of water displaced.

Man's Encroachment into the Oceans:

If we look at our coastal cities and towns, nearly all have built out into the ocean. Even cities around our great lakes have built out into their lake: i.e. Chicago, Cleveland, Toronto, Milwaukee, etc. Even my alma mater, Northwestern University in Evanston, Illinois, north of Chicago, built out into Lake Michigan. The displaced water went out the great lakes, Niagara

Climate Change: The Hoax of CO_2 Revealed

Falls and eventually out the Saint Lawrence Sea way into the Atlantic.

Of a bigger magnitude is the iconic San Francisco Financial District. Prior to the gold rush days that began in 1849, the entire district was part of San Francisco Bay.[16] Due to the demand for housing, to accommodate the gold rush miners on their way to the gold fields up the Yuba River, numerous northeast sailing ships found financial profits by serving as anchored hotels while charging inflated prices.[17] Thousands of these wooden ships lay anchored side by side for several years until the gold rush subsided. The city leaders at that time realized the financial value of the bay and soon land filled it so that tall skyscrapers exist there today. So, where did the water go that was 20 to 30 feet deep at that time? It was merely displaced further out into the ocean, raising its level slightly.

Another displacement of water previously unrecognized but also due to city encroachment was found shortly after our 9-11-2001 terrorist attacks.[18] When excavating the rubble of the twin towers an old British taxi ship was found beneath the level of the

Dr. Robert Marx

towers (Fig 3-8). The ship dated to 1754 and was used to taxi captured colonists to a larger British torture ship anchored further out in New York Harbor. The 9-11 towers were 500 yards inland from present day New York Harbor. This finding illustrates how New York, like all the other coastal cities, built out into their waterway displacing the sea level upward.

Figure 3-8

Fig 3-8 *The remnants of a 1754 British ship anchored in what was then New York Harbor was found beneath the 9-11 Trade Center rubble 500 yards inland from present day New York Harbor.*

Climate Change: The Hoax of CO_2 Revealed

One of the more visible man-made displacements of ocean water are the nine military islands built by China in the South China Sea.[19] Most were built between December 2013 and October 2015 and covered more than 3,000 acres that was once a living coral reef (Fig 3-9). Each individual encroachment may be small, relevant to the vastness of the oceans but adding them all together accounts for a sea level rise to be taken seriously.

Figure 3-9

Fig 3-9 *Aerial view of one of nine military islands built into the South China Sea by the Communist Party of China (CPC).*

Dr. Robert Marx

Natural Islands:

We need to realize that island chains such as the nine major islands of the Hawaiian chain as well the 13 major islands of the Galapagos chain were formed millions of years ago by undersea volcanic activity.[20] However, this activity continues today. In the Hawaiian island chain, the Big Island's two volcanoes recently erupted. Kilauea erupted on January 5, 2023[21] and its Big Sister volcano Mauna Loa on November 27, 2022.[21] In each, lava flowed out to the ocean boiling water as it entered and displacing water as it extended the size of each island. These two volcanoes also emitted an incalculable amount of CO_2. Please Note: The National Oceanic and Atmospheric Administration (NOAA) measures atmospheric CO_2 at their Mauna Loa facility atop the volcano perimeter. Add to this are the numerous other volcanoes emitting CO_2. Those near the water such as Krakatoa which erupted June 9, 2023 in Indonesia poured their lava into the ocean to displace even more water while also emitting CO_2.

Climate Change: The Hoax of CO_2 Revealed

Volcanoes: Today, there are more than 500 active volcanoes. When a volcano erupts, it spews out CO_2, sulfur, and methane, as well as an immense amount of particulate dust called ash. This dust is also comprised of gas and mineral contaminants, such as nickel and iron. In earth's history, the most massive extinction known occurred 200 million years ago due to excessive and repetitive volcanic eruptions in what is now known as Siberia.[22] This is referred to as the Permian extinction.

Scientists estimate that over 90 percent of the species on the earth at that time became extinct during its aftermath, clearing the path for the dinosaurs that became the dominant life form until their own extinction from the K-2 asteroid strike 65 million years ago.[23] Experts believe that the long-lasting dust cloud of fire and ash from each of these two separate events more than 185 million years apart covered animals and plants worldwide and that the dust cloud blocked out the sun, creating a prolonged "winter." Those who lived during the Cold War were warned of just such a winter, called the "nuclear winter," caused by repetitive nuclear blasts kicking up a large volume of particles into the atmosphere and blocking out the sun's rays for years, decades and perhaps centuries. Therefore, we should not underestimate the contributions of volcanoes and particulate matter to cause climate change either way.

In just the past 100 years, there have been over 100 minor and 12 major volcanic eruptions, not counting vents producing geysers and heated pools such as those in Yellowstone National Park, Iceland, and northern Japan. Today's volcanic events represent a quiescent time in earth's volcanic activity. One should note that the inner core of the earth is a molten pool in which certain openings in the tectonic plates allow heated vents, developing volcanoes, or actual volcanoes to occur.[24] Most noteworthy and active today are the Hawaiian Islands chain, the Galapagos Islands chain, the Mid-Atlantic Ridge, and the well-known "Pacific Rim of Fire," which stretches across the western coasts of South and North America, across the Alaskan-Bering straits area and down the Eastern Coast of Siberia, Korea and Japan as well as the Philippines and Indonesia.

Climate Change: The Hoax of CO_2 Revealed

Scientist have labeled the ten most active and arguably the potentially most destructive volcanoes in order:

1. Yellowstone: Right in our own backyard. If you live in the U.S. The Old Faithful geyser should be taken as more of a warning sign than a tourist attraction.

2. Mt. Vesuvius: If you visit Pompeii, Italy, which received the brunt of the Mt. Vesuvius first eruption in 79 AD, you will get a glimpse of the effect of volcanic ash as entire persons and their pets are seen encased in ash and preserved for us to see today. A second eruption is predicted at some time in the near future.

3. Popocatepetl: Located in Central México just 43 miles southeast of México City. This volcano continues to be active and has been heating up for decades. Its peak at 17,802 feet (3.4 miles) had a full glacier on it for centuries but melted as of 2001.

4. Sakurajima: Located in the southern Japanese island chain. This volcano's most recent eruptions occurred in 1914 and 2019 and continues to emit smoke and ash today. It is predicted to undergo a major eruption in the next 30 years.

5. Galeras: Located in the Southwestern area of Columbia, South America. This volcano has been continuously active with 33 eruptions since 1535, the most recent one in 2010.

6. Mt. Merapi: Located in Indonesia. It is the most active of the 129 active volcanoes in Indonesia. It erupted twice on June 21, 2020, which was preceded by a more deadly eruption in 2019 killing more than 300 people and sending a large cloud of ash four miles into the sky.

7. Mt. Nyiragongo: Located in the Virunga National Park within the Democratic Republic of the Congo. Mt. Nyiragongo is the most active volcano in Africa and houses an enormous lava lake within its crater.

Climate Change: The Hoax of CO_2 Revealed

8. Ulawun: Located on the Island of New Britain in Papua, New Guinea. This volcano is noted for its 22 recorded eruptions since 1700, the last one as recently as June 26, 2019.

9. Taal Volcano: Located just 51 miles South of Manila, the capital of the Philippines. This volcano is known for its violent eruptions, the last of which occurred on January 12, 2020.

10. Mauna Loa: Located on the "Big Island" of Hawaii and is actually the creator of that island. Mauna Loa is considered the second largest volcano on earth. Currently continually active, it is expected to undergo a full-eruption in the near future as its sister volcano Kilauea did in 2018.

The overall impact of an erupting volcano like the recent Kilauea eruption on Hawaii's Big Island is more than the dust cloud and CO_2. The impact on the climate is both controversial and immeasurable. That is, some scientists have claimed that the CO_2 given off in a single eruption is 10,000 times more than humankind has contributed to the atmosphere since we emerged from the caves.[25] Others relate that the

Dr. Robert Marx

CO_2 given off in an eruption is insignificant compared to the estimated 22 billion tons given off yearly by modern industry and our automobiles.[26] The problem with this controversy is that both are conjectures that are unverifiable. You cannot measure the amount of CO_2 produced by a venting pre-eruption Volcano or from its actual eruption. You also cannot measure the amount of CO_2 produced by cars, trucks and industry. Suffice it to say, a single volcanic eruption is preceded by the particulate matter in its smoke and venting for days and weeks prior to the main eruption and the CO_2, SO_2, and ash given off during the eruption is a very significant amount.

However, this topic exposes the repeated problem associated with governmental agencies and certain media using their given credibility to mislead the public. The argument put forth by NOAA, NASA, the DNC, Reuters and the AP etc., each claim volcanic CO_2 is insignificant because volcanic activity and eruptions are few and short term so that they emit less CO_2 than humankind. They seem to be ignorant of or purposely ignore the aforementioned hydrothermal vents in the Mid-Atlantic ridge

Climate Change: The Hoax of CO_2 Revealed

discussed earlier in this chapter. This ridge is a continuous outpouring of 700 degree Fahrenheit water and CO_2 constantly, every day, every year over a 300 to 600 mile width and 12,000 mile long range. Fake news and fake science exposed once again.

The process of undersea vents and volcanoes is also much more significant than realized as exampled by the new island that unexpectedly formed off the coast of Japan in October of 2023 [27] from an undersea vent and the completely under water eruption of the Tonga volcano in the South Pacific on December 20, 2021 and then again in 2023.[28] Actually, the Tonga volcano was under 500 feet of water. These unrecognized sources of heat, CO_2 and the additional land mass that is created unrelated to fossil fuels burning and man's production of CO_2 is something unaccounted for in the climate change models we are expected to accept today.

To further add to the contribution from undersea vents and volcanoes is the 2010 discovery of the mid Atlantic ridge. This ridge extends from Iceland 12,000 miles south midway between the European-African continents and the North and South American

Continents (Fig 3-10).[29] It varies between 300 miles and 600 miles wide and contains numerous heated vents. These hydrothermal vents continuously pour out superheated gases, one of which is CO_2 gas and molten rock between 575- and 700-degrees Fahrenheit (Fig 3-11).[30]

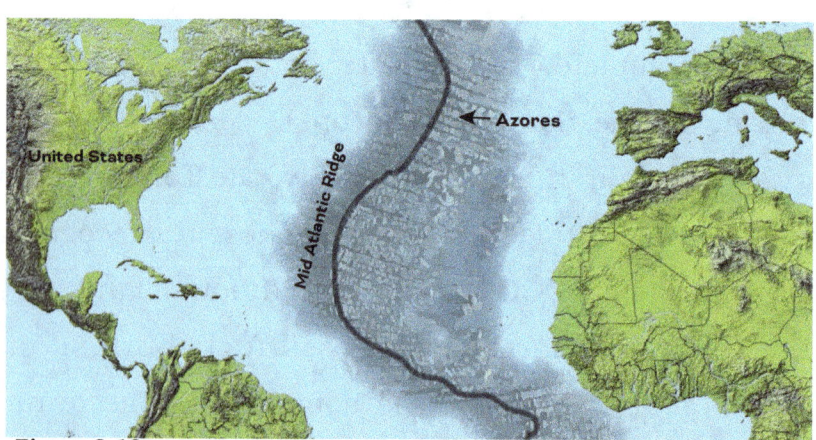

Figure 3-10

Fig 3-10 *The mid-Atlantic ridge is a 12,000 mile long by 300 to600 mile ridge midway between the Americas and the European and African continents. It has numerous hydrothermal vents and continuous volcanic activity.*

Climate Change: The Hoax of CO₂ Revealed

Figure 3-11

Fig 3-11 Hydrothermal vents in the Mid Atlantic Ridge continuously emit 575 degrees to 700-degree Fahrenheit magma and CO_2 into the surrounding water.

These vents were so numerous and impressive that the chief scientist from the expedition of the Marcum Center for Marine Environmental Sciences at the Max Planck Institute for Marine Biology in Bremen, Germany made the following salient observation "This could change our understanding of the contributions of hydrothermal activity to the thermal budget of the oceans." Indeed, the heat of the gases at 575-to-700-degree and the fact that one of the gases is CO_2 underscores that our atmospheric CO_2 is rising from our oceans and from other natural sources much more than from fossil fuels.

The Oceans are Getting Shallower:

This seemingly paradoxical statement is actually true. One only needs to look at an aerial view of every major river to see a plume of particulate matter extending out to sea 50 to 100 miles (Fig 3-12). This is seen from the Mississippi, the Columbia, the Amazon rivers as well as the many tributaries in Alaska, the Ganges and the Nile rivers plus many more. The particulate matter flowing out from these rivers is on the order of tons per second.[31] To underscore the magnitude of this particulate settlement, one needs only to look at two sunken Civil War ships. The union Iron Clad Monitor was found 16 miles off Cape Hatteras, North Carolina in 230 feet of water encrusted in Barnacles and sponges in addition to several inches of sediment.[32] More dramatic is the story of the Confederate States of America ship the Hunley known as the first submersible to sink an enemy ship. It sank the USS Housatonic on February 17, 1864 in Charleston

Climate Change: The Hoax of CO₂ Revealed

Harbor.[33] However, after torpedoing the Housatonic, it almost immediately sank. It sank in 30 to 40 feet of water, but when a diver discovered it in 1994, it was completely buried in six feet of silt that had covered it during the past 131 years.[33]

Figure 3-12

Fig 3-12 This space view of the Mississippi River Delta shows the expansive plume of runoff silt out into the Gulf of Mexico and stretching for over 100 miles down current.

The Earth is Accumulating More Water:

While the earth's surface today is 70% water, it was not always that way. When the earth formed 4.57 billion years ago the elements of hydrogen (H_2),

Carbon C, and methane (CH_4) were present but only a small amount of oxygen O_2 was present to form water (H_2O). Even carbon dioxide (CO_2) could not easily form due to the scarcity of oxygen. It was not until primitive algae and other forms of early plant life evolved that oxygen levels increased sufficiently to react with Hydrogen to form water. Comets, meteors and asteroids added more water to this initial start in the form of ice. Over hundreds of millions and billions of years our oceans formed. This process and such impacts are continuing today. Although earth has not received a major meteor or asteroid impact in the last one hundred years, the 2013 meteor impact in Chelyabinsk Russia is noteworthy.[34]

It is known that 37,000 to 78,000 tons of meteor/meteorites strike the earth each year.[35] Most are just dust like but still carry a small amount of water or give off the water in the form of steam vapor as they burn up coming into our atmosphere. Each of these space particles contain water and are called carbonaceous chondrites. Some are larger (0.35 ounces to 2.2 pounds) and carry more water. The Chelyabinsk meteor was 66 feet in diameter and

Climate Change: The Hoax of CO_2 Revealed

entered our atmosphere at 42,690 miles per hour.[34] The water content it added when it exploded was documented on film as a water vapor trail (Fig 3-13). Additionally, the fact that it exploded, leaving little solid material confirms that this meteor was almost solid ice. The physics of its explosions short of an impact into the ground is similar to that of popcorn where the expanding heated water vapor pressure exceeds the structural integrity of the husk causing an explosion that pops the kernel.

The simple fact is the earth has gained volumes of water by this process since it was formed and continuously does so slowly today. As a planet, we slowly gain water but never really lose it.

Dr. Robert Marx

Figure 3-13

Fig 3-13 *Vapor trail indicating the large water content of the Chelyabinsk meteor consistent with most meteors.*

Climate Change: The Hoax of CO_2 Revealed

References
Chapter 3

1. Kruger T. Discovering the Ice Ages. International reception and consequence for a historical understanding of climate (German edition Basel 2008) Leiden 2013, p47, pp78-83, pp150-151
2. Kim Ann Zimmerman-Live Science Aug. 29, 2017. Pleistocene Epoch: Facts about the last ice age. https://www.livescience.com/ 40311-pleistocene-epoch.html.
3. Lawler A. Did modern humans travel out of Africa via Arabia. Science 2011, 331 (6016):387. Bibcode 2011 Sci...33 1 ...387L[.doi:10. 1 126/science.331.6016.387[.PMID 21273459{.
4. https://news.stony brook.edu>Newsroom. SBU News: First Modern Humans Arrived in Europe Earlier than Previously Known
5. United States Geologic Survey (.gov.). The coastline of the eastern US changes...slowly. https://www.usgs.gov>media>images>coasUine-

easter...

6. "The Doggerland Project", University of Exeter Department of Archeology
7. White MJ. Things to do in Doggerland when you're dead. Surviving 0153 at the Northwestern Most Fringe of Middle Paleolithic Europe. Nov. 2006 (PDF) World Archaeology. 38(4):547-575 doi.10.1080/00438240600963031[.S2CID 51729868]
8. O'Dea A, Lessios H, Coates A, et al. Formation of the Isthmus of Panama. Sci Adv 2016, Aug;2(8):e1600883
9. Bermuda's Fish. Government of Bermuda. Department of Environment and Natural Resources.environment.hm/fish
10. Humboldt Current-Wikipedia https://en.wikipedia. Org>wiki>Humboldt_Current
11. Smith DG. Glacial Lake McConnell: Paleogeography, age, duration, and associated river deltas Mackenzie River basin, Western Canada. Quaternary Science Reviews. 1994, 13 (9- 10):829-843. Bibiode: 1 994QSRV...

Climate Change: The Hoax of CO_2 Revealed

13.8295.doi.10.1016/0277-3791

12. Cree-Lake-Indigenous Saskatchewan encyclopedia University of Saskatchewan.https://teachingusask.ca?indigenous>import>cre

13. What is Glacial Isostatic Adjustment? National Ocean Service(.gov) https://oceanservice.noaa.gov>facts>glacialadjustments

14. How they were made. Wisconsin Sea Grant

    https://wwwseagrant.wisc.edu>resources> how they were made.

15. Levy, K. Rare petroglyphs in Michigan provide link to Native American past. Lansing State Journal. June 9, 2013.
    https://www.lansingstatejournal.com>2016/06/09 >m

16. Archimedes Principle and Buoyancy. BC Campus Pressbooks.

    https://pressbooks.bccampus.ca>chapter>archemedes...

17. Financial District, San Francisco. Wikipedia.

    https://en.wikipedia.org>wiki>financial district,

sa ...

18. Wooden Ship Unearthed at World Trade Center From Revolutionary War. National Geographic. https://www.nationalgeographic.com>history>article

19. China's Island Building in the South China Sea" Damage to the Marine Environment, Implications, and International Law. US China Economic and Security Review Commission (.gov) https://www.uscc.gov>research>chinas-island-bild...

20. History of Galapagos. Galapagos Conservancy https://www.galapagos.org>About Galapagos

21. Active volcanoes of Hawaii. United States Geological Survey (.gov). https://www.usgs.gov>observatories>hvo>active-v...

22. Permian extinction, facts and information. National Geographic. https://wwwnationalgeographic.com>science>article

23. K-T extinction/overview & facts. Britannica. https://www.britannica.com>science>K-7 extinction

Climate Change: The Hoax of CO_2 Revealed

24. What is a hydrothermal vent? - National Ocean Service. National Oceanic and Atmospheric Administration (.gov).
https: / /oceanservice.noaa.lgov>facts>vents
25. Volcanoes Dwarf Humans for CO_2 Emissions. NBC News June 27,2011
https://www.nbcnews.com>wbma43554668
26. Humans emit ore CO_2 than volcanoes. AP news https: //apnews.com>article>fact-check-volcanoes-co...
27. Lea, R. Satellites watch as Japan's new volcanic island continues to grow. Space.com.
https://www.sparse.com>satellites-Japan new-island-still-...
28. Pare S. Tonga volcano eruption was fueled by 2 merging chambers that are still brimming with magma. Live Science Dec, 19, 2023. https: / /www.livescience.com>planelearlh>volcanos
29. The Mid-Atlantic Ridge, UNESCO World Heritage Centre.
https: //www.whc.unesco.org>activities
30. Hydrothermal vents. Woods Hole Oceanographic Institution. https: / /www.whoi.edu>...

>hydrothennal vents

31. Mississippi River Basin/ Gulf of Mexico Nutrient Runoff Network
Info Bulletin. Sep. 11, 2023 National Oceanic and Atmospheric Administration (.gov) https://www.noaa.gov>sites>default>files (PDF)

32. USS Monitor Center cleans the inside of one of two engine rooms. The Civil War Project. https://civil-war-picket. blogspot.com>2020/ 02>uss...

33. H.L. Hunley Wreck (1864). Naval History and Heritage Command (.mil).
https://www.history.nayy.mil>ship--wrecksites>hl-hu ...

34. What was the Chelyabinsk meteor event? The planetary society
Feb. 15, 2023. https://www.planetary.org>articles

35. Meteorites on Earth: How many fall per year and why we don't see them? Iberdrola.
https://www.iberdola.com>innovation>meteorites-earth

Climate Change: The Hoax of CO_2 Revealed

Chapter 4

The Greenhouse Sleight of Hand

Those obsessed over CO_2 have advanced the notion that CO_2 is causing global warming by a greenhouse effect while ignoring the solar variations, the past warming periods unrelated to CO_2, the Milankovitch cycles, and the myriad of other data identifying the real reasons why earth is currently in another one of many known past history warming periods.

Actually, CO_2 and the other "greenhouse gases" i.e. H_2O, O_3, CH_4 do not act as does a real greenhouse. A real greenhouse has a floor, four walls and a ceiling as a completely closed system. Solar energy enters through the glass or clear plastic that makes up the four walls and the ceiling. The materials inside absorb light energy, heat up and radiate their heat into the airspace of the greenhouse. The heat is then trapped. This allows vegetables and other plants to grow in cold winter months when the outside temperature is too cold to otherwise support certain plants.

Although the greenhouse effect in controlling earth's atmosphere has been present for over 4 billion years, it is not a true greenhouse. That is, earth and the troposphere as well as the stratosphere are all connected as an open system. The purported greenhouse gases absorb infra-red rays and therefore absorb heat radiating from the surface and acts as a porous blanket not a sealed off roof. If it were not for the greenhouse gases, the earth would be much colder and would likely freeze over, similar to the snowball earth ice age of 850 to 630 million years ago. The actual greenhouse gases are beneficial, acting as a thermal regulator to earth's climate, particularly the stable climate humans have enjoyed and prospered from since the ending of the last ice age.[1]

The problem of the CO_2 obsessed is their failure to realize that the greenhouse effect is incomplete. This allows much of the heat energy from our solar rays to indeed escape into the stratosphere consistent with an open system greenhouse. In effect, the greenhouse

Climate Change: The Hoax of CO_2 Revealed

gases together act as a thermo-regulator. The CO_2 obsessed also do not seem to understand that H_2O in the form of humidity and actual clouds has over 30 times the greenhouse effect, as does CO_2.

Clouds in particular were not factored in the computer models of Hansen or the "Hockey Stick" computer model of Mann,[2] both of which have been proven wrong by their false predictions and scholarly articles by nongovernmental supported scientists.

The acey-deucey part of clouds is that on one hand they reflect the sun's rays back into space and create shade reducing the heat radiating from earth's surface. This produces a cooling effect. However, if clouds act as a greenhouse, it should be known that its effect is 30 times more than CO_2. Clouds have a strong warming effect. Most everyone who has lived through a cold winter has observed that a bright cloudless day in the dead of winter is colder than one with a gloomy overcast sky. This underscores the real nature of how our greenhouse gases work. They are not constant. There are days-weeks-months-years etc. when the greenhouse blanket is a full cover and times when it is less. This has occurred over the timeframe of the last 50 years when so much vilification of CO_2 has taken place while there hasn't been a significant increase in global temperatures.

As stated in chapter one, CO_2 also has a finite time in the atmosphere where it can act together with H_2O, CH_4, O_3 as a greenhouse gas. It may surprise some people to know that some CO_2 escapes into space, much becomes dissolved in our oceans and our lakes, some stays in our atmosphere for a limited time and some is taken up by plants for photosynthesis. That is, CO_2 does not accumulate in the atmosphere. It is in a dynamic equilibrium with these other reservoirs. It has a reported 3 ½ year life span in the atmosphere.[3]

The Absence of a Link Between CO_2 and Global Warming:

Our current atmosphere is composed of nitrogen (N_2) 78.03%, oxygen (O_2) 20.95%, Argon (Ar) 0.93%, CO_2 0.042 % and trace gases 0.048%. This represents an increase of 0.012% of CO_2 since 1940. This then represents a 2.3% increase in CO_2 over the past 84 years, yet, the earth's temperature has not increased

Climate Change: The Hoax of CO_2 Revealed

by 2.3%. In fact, accounting for all the cold spikes, warm spikes, El Niño's and La Niña's there has been no real increase in the average measured temperature.

The fact of the matter is that the political-media complex has invented and perpetuated a falsehood of a direct connection between CO_2 and the local spikes in temperature. A couple examples are worthy to recall.

Example 1:

A young Miami News reporter, reporting for a local news program, was reporting evidence of sea level rise while standing in a foot of water 100 yards from the shore. She lamented that global warming due to CO_2 caused this flooding and will only get worse unless drastic measures were undertaken soon. What she didn't know or failed to research is that it was the time of Miami's "King Tide" where the phases of the moon produce the highest tides of the year. More importantly and indicative of today's superficial media investigative research is that the area in which she was standing was underwater in 1930 before it was landfilled to extend the livable land of Miami Beach.

Example 2:

The November 23, 2023 edition of the Miami Hearld's Newspaper depicted an arborist from California taking a tree sample from a dead tree. The article went on to indicate that drought was the probable cause of the tree dying. Of course, the writer attributed the drought to CO_2 and global warming. It warned that continued tree deaths that are now occurring are due to global warming while the picture of the arborist showed healthy appearing trees in the background.

Apparently, the writer of this article never heard of the infamous "dust bowl" that ravaged Oklahoma, Kansas, Nebraska, and even the plains of Canada.[4] The "dust bowl" was a severe drought that began in 1930 (97 years before this article) and extended until 1939 (Fig 4-1).[5] During this time, crops were lost, lakes dried up, many people died, and many moved away. Imagine if a similar "dust bowl" occurred today. If it did, CO_2 would surely be blamed. Our

Climate Change: The Hoax of CO_2 Revealed

government would look to mandate immediate untested remedies while the media decries it blaming all of us for producing CO_2.

Figure 4-1

Fig 4-1 *Drought conditions in the United States from 1930 to 1939 known as the "Dust Bowl" produced real dust storms that resemble those produced in today's movies.*

Example 3:

Perhaps more sinister but even more important to expose comes from Laurie David the producer of AL Gore's movie the "Inconvenient Truth".[6] As pointed out in Chapter 1, the book an subsequent movie failed to identify that the rise of CO_2 came as a result of the global warming after the last ice age and from the other causes that I illustrated in Chapter two. Instead, the movie attributed the rise of CO_2 as the cause of global warming.[7] After the May 24, 2006 release of the "Inconvenient Truth", he published a children's book in 2007. The title was "The Down to Earth Guide to Global Warming" (Fig 4-2).[8] The book emphasized the disastrous consequences of CO_2 induced global warming from droughts to floods to hurricanes and starvation among others. The book featured a figure depicting CO_2 in the atmosphere and climate temperature during the past 650,000 years. The book braggingly claimed that it demonstrated "the line" between CO_2 and greenhouse gases and global warming. To make it appear that a CO_2 rise came before global warming, he switched the tables in the figure.[9] The original and correct labeling showed the temperature rise to precede the rise in CO_2.

Climate Change: The Hoax of CO_2 Revealed

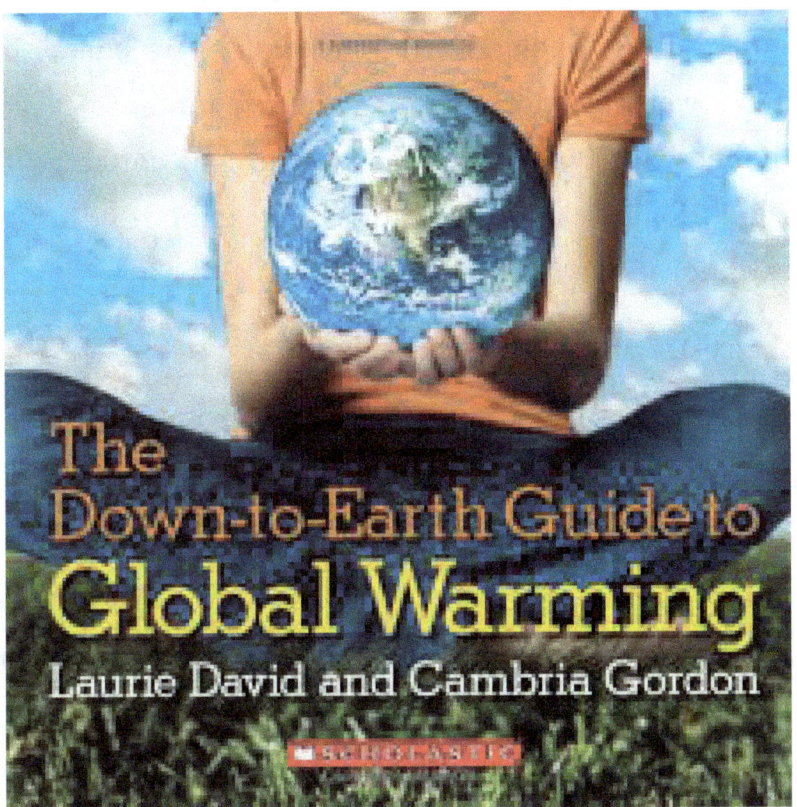

Figure 4-2

Fig 4-2 *The book "The Down to Earth Guide to Global Warming" targeted impressionable children and teenagers with false data and dire predictions that were later proven to be wrong.*

When confronted with this second falsehood by the science and public policy Institute, Laurie David dismissed it as just a "minor error".[9]

Dr. Robert Marx

This so-called minor error is not minor at all. It was a planned lie to mislead the reader. In this case, the readers were children and teenagers, just the target audience who the CO_2 obsessed want to convince to become loyal supporters of their climate change agenda for years to come.

What to Do With CO_2?

The hypocrisy and an avoidance of the real science that points to the actual causes of global warming continue at the highest levels. The Biden-Harris administration and some state governors joined the CO_2 obsessed and continually lobby for the "Green New Deal", carbon credits (cap and trade). Electric car mandates, and a carbon tax all of which are unnecessary and would not alter global warming one bit.

To expose their hypocrisy, one must ask why they are not promoting a concerted reforestation program or a program to re-seed coral reefs? Trees and corals are the most significant natural entities that take CO_2 out of the atmosphere. Instead, they focus only on fossil fuel production of CO2 while also ignoring the

Climate Change: The Hoax of CO_2 Revealed

natural CO_2 production from volcanoes, hydrothermal vents, the release of dissolved CO_2 from our oceans etc. Instead, they seem to be only focused on fossil fuel emittance of CO_2. Ask yourself why.

Trees absorb CO_2 and emit O2 in the classic photosynthesis process. However, the earth is estimated to have had 10 trillion trees when Columbus landed in 1492.[10] It now has an estimated 3.1 trillion trees; a loss of 6.9 trillion trees over the past 532 years.[11] This tree loss and the loss of the ability to scrub CO_2 from the atmosphere has been going on steadily commensurate with human expansion and population growth. We should recall forests once stood where New York, Boston, Chicago, Los Angeles, San Francisco, Munich, Paris and most every large city and small town stand today. We forget about the past and now look to the rapid deforestation occurring in the Amazon River Basin where trees are lost at the rate of one acre per second.[12] Add to that alarming figure, the rate of deforestation or "land clearing" to accommodate population growth around the world is at a rate of one and one half acres per second.[12]

Dr. Robert Marx

When trees are lost, their ability to retain water that would otherwise enter the oceans is also lost. About 30 to 50 percent of a tree's weight is water.[13] Therefore, a 30-foot tree will contain 150 to 200 gallons of water within its roots, trunk, and branches. This extrapolates to 620 trillion gallons of water that is kept out of our oceans in today's trees and 1,380 trillion gallons (1.3 quadrillion gallons) added to our oceans via tree loss in the last 534 years. Sea level rise is added to by every lost tree.

If you feel that CO_2 is really the problem and want to combat sea level rise, reforestation represents the most direct and practical way to deal with it. Is our government or any other global government taking on this challenge? The answer is sadly NO.

To illustrate the proven value of reforestation is the example set forth after the post St. Helens volcano eruption in 1980. Volunteers and the Weyerhaeuser Corporation set up a unique experiment that turned out to be a template for reforestation. The massive eruption of Mount St. Helens devastated and laid bare the landscape for miles around it. Community volunteers and Weyerhaeuser Corporation cleared

Climate Change: The Hoax of CO_2 Revealed

and removed enough dead trees to build 85,000 three-bedroom homes in 45,000 acres of the 90,000 acres that was laid bare.[14] They then planted 18 million sapling trees over the next three years. The remaining 45,000 acres were left to recover and reforest by nature. Today, the community-Weyerhaeuser project sports a complete reforestation with 60-foot-tall trees, the return of wildlife and the prevention of soil erosion.[14] It also holds five billion gallons of water out of our oceans and has taken an immeasurable amount of CO_2 out of the atmosphere. The 45,000 acres left to nature still remains mostly barren with small shrubs, bushes and little wildlife. Note, neither I nor any of my family have any financial or personal connection with the Weyerhaeuser Corporation. Credit is advanced where credit is due.

But what kind of credit can we give our President, Vice President, certain Governors, The Environmental Protection Agency (EPA), The Secretary of the Interior? Why have they not seen this as a positive example and many of the other small scale successful projects to restore our forests? The answer is that it is easier to tax, to regulate, and to mandate. Reforestation takes leadership, organization, commitment, and hard work. Those leaders have none of these qualities.

Coral Reefs:

Coral, in addition to trees, is one of the major organisms that removes CO_2 from the environment. While trees use CO_2 for photosynthesis to build bark, leaves, and the very wood of the tree, coral polyps take up CO_2 to synthesize their calcium carbonate ($CaCO_3$) outer shells. Coral reefs, such as the Great Barrier Reef, John Pennekamp State Park, the Maldives and the Bahamian reefs etc. represent a vast subsurface forest (Fig 4-3).

Climate Change: The Hoax of CO$_2$ Revealed

Figure 4-3

Fig 4-3 *Coral reefs represent a vast undersea forest that removes dissolved CO$_2$ from the largest known reservoir of CO$_2$: our oceans.*

It is well to remember that coral is an animal, not a plant. It is a polyp that encases itself in a calcium carbonate shell and fixes itself to a convenient surface. It has a unique relationship to a mutualistic algae species, zooxanthellae which provides oxygen to the polyp and gives it its color.[15] The polyp provides CO$_2$ and other nutrients for the algae to photosynthesize so as to maintain its own existence and retain the color of the coral. Coral is fragile and is noted to be in decline in many parts of the world.

Dr. Robert Marx

Of course, the CO_2 obsessed attribute the decline totally to global warming which on its face would seem plausible due to the fact that dead coral reefs have a ghostly white appearance called "bleaching". However, when the zooxanthellae algae dies due to a diminished or absent light energy to support its photosynthesis, the coral polyp also dies from a lack of the oxygen and the nutrients that the photosynthesis process releases. When the dependent algae responsible for the color dies, the now hollow $CaCO_3$ encasement of the coral polyp remains and the structure turns the natural color of $CaCO_3$ which is white (Fig 4-4A,B). The death of coral and the decline of coral reefs is not due to CO_2 or global warming. It is due to the increasing turbidity of the water that most all who have been near an ocean can attest to.

Climate Change: The Hoax of CO_2 Revealed

Figure 4-4A

Figure 4-4B

Fig 4-4 (A,B) *Bleached coral is not bleached by the sun as some have advanced. Instead, it is caused by the death of the mutualistic algae zooxanthellae which is the source of the color. Note! The zooxanthellae die due to turbid water blocking out the sun's light energy they need to survive.*

Dr. Robert Marx

Attributing coral decline to global warming is false by sheer logic and direct observations. That is, if coral decline is due to increased ocean temperatures, one would expect the pattern of coral death would be more prominent in the warmer waters. Yet, it is not. In fact, the reefs most resistant to coral decline are the southern areas with the warmest waters. One would also expect the northern areas or the areas on the edge between coral reefs and water too cool to support coral to begin to sprout coral. This is not happening.

So, what is actually causing coral decline?

Some have suggested that increased ocean acidity due to CO_2 is the cause of coral bleaching because of the chemical equation of $H_2O+CO_2=HCO_3^-+H^+$ with HCO_3^- being a weak base and H^+ a strong acid. Yet the measured pH of our oceans continues to be steady between 8.1 and 8.2 which is well on the basic side of neutrality between an acid and a base pH = 7.0.[16]

The correct answer is in front of our very eyes. It is our obscured vision from murky water. Coral requires direct sunlight to continue its mutualistic relationship with the zooxanthellae algae which is a light requiring plant. This is why most corals are seen in water 100 feet or less and in clear water. The pattern of coral reef loss most directly relates to water clarity and is seen mostly around coastal cities and areas of particulate run off. As a skin diver myself, I have noticed the increased turbidity of the water and loss of portions of a coral reef off of Key West, Florida. This was emphasized and accelerated when cruise ships began docking in the Key West Harbor (Fig 4-5). After several years, its negative effects on the coral as well as on skin diving, fishing and pleasure boating, prompted the voters to pass an amendment to the city charter prohibiting larger ships, limiting docking, and approving only ships with a superior environmental record. The vote, in November of 2020, passed with a 61% to 81% approval. Yet today larger cruise ships frequently dock in Key West Harbor.[17] It should be noted that when the COVID-19 caused a prohibition of cruise liners docking in Key West Harbor the water clarity noticeably improved but declined again upon the resumption of Cruise Line activity.

While we scientists support and look to robust controlled studies, direct observation remains an important and verifiable research tool.

Fig 4-5 Cruise ship docking at Duval pier in Key West harbor producing a large cloud of silt particles that currents take out to cloud the waters and settle on nearby coral reefs.

Climate Change: The Hoax of CO_2 Revealed

References
Chapter 4

1. Singer SF. Hot Talk Cold Science. Independent Institute, Oakland CA 2021, ppl53-160
2. Montford AW. The Hockey Stuck Illusion: Climategate and the Corruption of Science. London Stacey International 2010
3. Singer SF, Avery OT. Unstoppable Global Warming: Every 1,500 years updated and expanded edition. Oct. 22, 2007. Rowman and Littlefield Publishers.
4. Timeline: The Dust Bowl! American Experience Official Site. PBS. https://www.pbs.org>dustbowl-surving-dust bowl
5. Dust Bowl: Causes, Definitions & Years/History. History Channel. https://www.history.com>topics>great depression
6. Gore A. An Inconvenient Truth: The Planetary Emergence of Global Warming. Rodale Books, May 26, 2006

7. An Inconvenient Truth: Filmmaker Davis Guggenheim IMDB. https://www.imbd.com>title
8. Gordon C, Laurie D. The Down to Earth Guide to Global Warming. Google Books 2007
9. Singer SF: Hot Talk cold Science. Independent Institute Publisher 2021, pp90-91
10. Connor S: Earth has lost more than half its trees since humans began cutting them down. Independent. Co Sep. 2, 2005
11. Marx RE. Climate change The Real Story. Dorrance Publishing, Pittsburgh PA 2022 p58
12. Sandy M. Why is the Amazon Rain Forrest Disappearing? Time Magazine. https://time.com>amazon-rainforest-disappearing
13. Water and Forests. US Forest Service (.gov) https://www.fs.usda.ogy>stelprdb52698 13 (PDF)
14. 25 years after Cataclysmic Eruption. Mount ST. Helens Forest Healthy and Productive: Witness the Return of the Forest at Weyerhaeuser's Forest Learning Center. Feb. 16,

Climate Change: The Hoax of CO_2 Revealed

003. Investor.Weyerhaeuser.com/200-02-16-25-years-after cataclysmic-eruption-mount-st-helens-forest-health-and-productive-witness-the-return-of-the-forest-at-Weyerhaeusers- Forest-Learning- Center

15. Why Corals are so Colorful? Woods Hole Oceanographic Institution. https://www.whoi.edu>knowyourocean>didyouknow.

16. Loaiciga HA. Modern-age Buildup of CO_2 and its Affects on Sea Water Acidity and Salinity. Geophysical Research 2006 Letters33:Ll0605, https://doi.org/ 10.1029 /2006GL026305

17. Key West Leaders Want Cruise Ship Limits. Miami Herald July 19, 2021. https: //www.miamiherald.com>local>article252759513

Dr. Robert Marx

Chapter 5

Fake Science

The political media complex has been the main purveyors of fake news, mandates, and regulations vilifying CO_2 as the sole cause of global warming. Their motives have been exposed as a push for the so-called Green New Deal, taxes, votes, power, retention of power, and most importantly control of its citizens. While they focus on this, they are not really doing anything to combat the climate change they decry. The media are in it to gain political favor, vie for a Pulitzer Prize or one of the many other lower-level awards they give to themselves. They also seek to gain favor and support from certain politicians to gain exclusive interviews and more viewers from their committed base which translates into more sponsors, more notoriety and of course more money.

But what about scientists? What could possibly be their motives? As an oral and maxillofacial surgeon

Climate Change: The Hoax of CO_2 Revealed

and a research scientist myself, I have firsthand insight into this. Simply stated, it is funding. The first goal for any researcher is to "get funded". The second goal is to renew funding. Researchers do not produce a salable product while they apply for grants or when they do the research that may eventually result in a salable product. They are totally dependent on the money they receive from their grant. I have personally witnessed and experienced worthwhile and important research proposals wither on the vine due to lack of funding related to a not politically correct study. Most researchers in universities and organizations such as NASA, NOAA, the IPCC, the UN, and the EPA have a small or nonexistent base salary. Some are totally dependent on their research grants for their livelihood. Therefore, while they may prefer to pursue a study other than one that is looked upon favorably by their parent organizations, they are strongly incentivized to pursue one that is aligned with the funding organization's ideology. This virtually guarantees it to be funded.

Examples from medicine include HIV studies during the 1980s AIDS epidemic. HIV research was heavily funded even for studies only remotely related to AIDS. During the COVID-19 epidemic, vaccine and viral studies were routinely funded even those of dubious scientific value.

Today, climate change, specifically global warming studies are predictably and often heavily funded with a bias that supports CO_2 based global warming and apocalyptic outcomes. Examples, of exposed fraudulent global warming studies beyond the exposed fraud of Al Gore's "The Inconvenient Truth"[1] and Laurice David's switching the graphs on his children's book "The Down to Earth Guide to Global Warming"[2] are:

Example 1: The Hockey Stick Graph Illusion:

In 1998, Michael Mann, a PhD in geology and geophysics, published a graph depicting a dramatic vertical upswing in temperatures on the scale 0.75 degrees Celsius (1.35 degrees Fahrenheit) between 1961-1990 compared to temperatures from 1,000 AD forward.[3] This, of course, was met with acceptance and glee as well as a "we told you so" attitude by the UN's Intergovernmental Panel on Climate Change

Climate Change: The Hoax of CO_2 Revealed

(IPCC).[4] This was just what they wanted to hear, a scientific smoking gun linking mankind's production of CO_2 turning the world into an oven. It became the center focus of the UN's IPPC Third Assessment Report (AR-3).[4] This graph known as the "Hockey Stick Graph" was, of course, prominently displayed in Al Gore's "The Inconvenient Truth". (Fig 5-1)

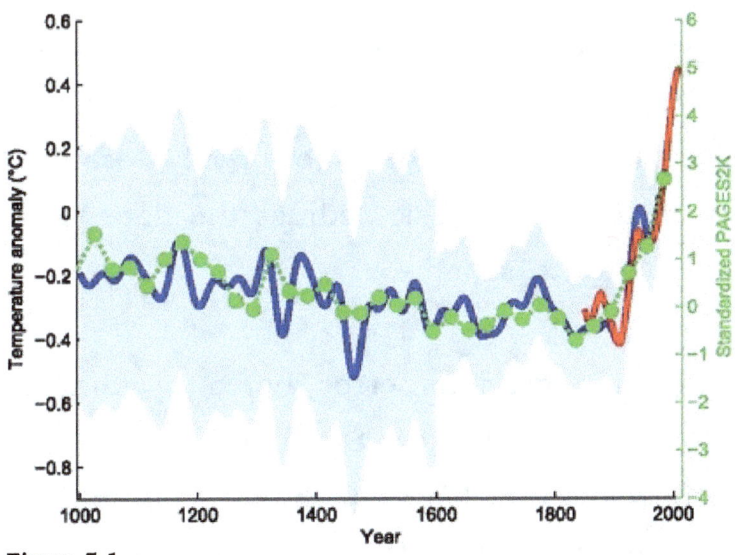

Figure 5-1

Fig 5-1 *The Hockey Stick Graph which falsely depicted a rapid and significant upswing in earth's temperature*

So, what was wrong or false about the Hockey Stick Graph? S. Fred Singer, with David R. Legates and Anthony R Lupo in their book Hot Talk, Cold Science relate that the first error was either sloppy research or intentional deception. That is, temperature records from 1900 to 1980 were substituted to falsely represent the actual temperatures of 1000 to 1980. This eliminated the troublesome medieval warming period of 1100 - 1400 AD which was warmer than that of 1998 and that of today. The graph also eliminated the colder temperatures of the "Little Ice Age" between 1400 to 1850. Singer and Legates noted that Mann put in over 900 years of stable global temperatures which did not really occur. This amounted to the classic "adjust the data to fit the result you want" scheme.[5,6]

The second egregious error noted by Singer and Legates is the Hockey Stick graph itself. These authors relate that Mann at the last minute changed his temperature methodology to capture the large El Niño effect of 1998 which significantly skewed the data and which subsided in 1999 and after. In doing so, he biased his data similar to one pronouncing global warming on one hot summer day or one denying global warming based on a single bitter cold winter day. Mann also did not include the colder annual temperatures from 1978-1997 that preceded the 1998 El Niño effect. This is an example of overt data manipulation that was also exposed by McIntyre and McKitrick in a publication that appeared in Energy & Environment and peer reviewed by the World Data Center for Paleophytology.[7] In using Mann's own methodologies without deleting the inconvenient data that would be contrary to Mann's preconceived outcomes, they found that temperatures prior to 1400 exceeded any values Mann attributed to the 20th century. Note: Mann and his coauthors published a correction in Nature admitting to their errors. However, the damage was done.[8] This was what the UN, IPCC and Al Gore needed to move on with a hoax that has been accepted and promoted by the political media complex and never made available to us average citizens.

Example 2: Corruption of the Peer Review Process

While the scientific peer review process can be boring and insignificant to the average citizen, it is an exceedingly important process to eliminate bias and false data. It seems that several members of the UN's

Dr. Robert Marx

IPPC are on many science journals, editorial boards and are journal editors themselves. The corruption and therefore wrong conclusions come from only selective data being included. A quote from an author who submitted a counter argument critical of an article to be published documents this quote from the journal editor, I do "NOT" want to "show the most recent radiosonde (balloon results)".[9] This data did not appear in the final manuscript published in the International Journal of Climatology.[9] Had it been published the scientific community would have found the original author's conclusions to be wrong.

Further corruption comes from letters to the Editor critical of a scientific paper. Science journals often choose not to publish a Letter to the Editor that points out the flaws in one of their already published articles. They also will preferentially publish a Letter to the Editor that is favorable to the ideology of the Editor-in-Chief. Additionally, Editors sometimes publish a letter critical of a paper without allowing the original author the opportunity to answer the criticisms. One glaring example comes from Tom

Climate Change: The Hoax of CO_2 Revealed

Osborne, a member of the editorial board of the International Journal of Climatology, in emails to the Journal Editor Glenn McGregor. "Correct the scientific record" and identify in advance "reviewers who are both suitable and available" perhaps including "someone on the email list you have been using".[9] The word suitable refers to reviewers who are sympathetic to the global warming advocacy of these two. Osborne further goes on to write that "McGregor" may be able to hold back the hard copy appearance of Douglass et. al. (a rebuttal to an author supporting global warming).[9]

The point of all this is to emphasize that the scientific literature has been tipped in favor of a global warming crisis both advancing false data and ignoring or suppressing contrary data. The triad of politicians, the media and biased scientists have made an incorrect concept seem legitimate. Have scientists and/or the UN been wrong before? Heck yes! The following are just a few examples that the public will recognize.

Example 1:

Dr. Robert Marx

The COVID-19 pandemic taught the public that the World Health Organization and the UN are not only biased but corrupt as well. The public should recall that the UN as well as Dr. Anthony Fauci told us that the COVID-19 virus came from nature not the Wuhan Lab in China and that the virus was not transmissible from one human to another.[10] I think we all know how proven wrong both proclamations were. It is also good to remember the laudable praise given to China for their transparency and management of the COVID-19 pandemic by the UN Secretary General, Antonio Guterres.

Example 2:

Scientists noted an increase in melanomas, a most deadly skin cancer, often called "black mole cancer" since 1935.[11] They attributed it to ozone depletion from fluorocarbons commonly found in hair sprays and refrigerants. They noted a particular increase in the mid 1960s and warned of epidemic levels predicted by computer models. In fact, laws were put into place restricting fluorocarbons to ward off the predicted epidemic of cutaneous melanomas. Well, like global warming, melanomas have been on a slow

Climate Change: The Hoax of CO$_2$ Revealed

but steady rise since 1960, not the exponential rise predicted by the models. What went wrong? The answer once again is the same fact as the computer models predicting the Hockey Stick Jump in atmospheric temperatures. That is, wrong information in – more than wrong information out. The slow increase in cutaneous melanomas arises from the advent and common usage of air conditioning that became available to the public in the 1960s. This caused an influx of fair skinned white northerners often with red or blond hair (the most vulnerable population for cutaneous melanomas) to move to the sunbelt states. The link to this population shift and the rate of this movement was not factored into the modeling. The link of cutaneous melanomas directly correlates to the increase of white skinned individuals in the south. Today, the ozone in the atmosphere has stabilized even though the current TRESemme hair spray and others hair sprays contain fluorocarbon 152A. Curious, isn't it.

Example 3:

In my own field of surgery, the World Health Organization (WHO) changed the designation of a familiar jaw cyst to that of a tumor. This was based on a gene mutation suggestive of a tumor called the PTCH gene mutation.[12] Without factoring in the actual behavior of this cyst clinically, they went ahead and changed the designation name to that of a tumor with the full authority of the WHO. This led to more aggressive surgeries including partial jaw amputations that had not been necessary in the past. Soon, thereafter, this same gene mutation was found in other cysts and even inflammatory conditions. The WHO under great pressure from the oral and maxillofacial surgeons and oral and maxillofacial pathologists as well as their own embarrassment throughout the medical world, rescinded their previous tumor designation and returned to the time proven designation of a cyst.[13]

The simple lesson learned from these examples is that there is always one thing that is more reliable than a computer model, or false data, or even a randomized prospective double blinded placebo-controlled study and that is the track record. The track record of global warming shows that it has nothing to do with human's production of CO_2 as

Climate Change: The Hoax of CO_2 Revealed

proven by earth's past history, satellite weather recordings, ice cores, deep-sea cores, solar radiation studies, apocalyptic predictions that have not come to pass, the knowledge of natural CO_2 production, and reservoirs of CO_2 among a host of others.

Example 4: The Y2K Event

Everyone over 30 years of age should recall the Y2K (year 2000) hysteria (Fig 5-2). Those under 30 years of age should look it up for a good lesson and laugh. In the years (1998 and 1999) preceding the turn of the century, a near panic emerged and built. It seems, we, as a society, became too dependent on computers, especially those that controlled the services which we relied upon such as gasoline pumps, water pumps, banking and accounting, and even the newest thing, the internet. It was conjectured that the timestamps on these devices were only programmed to the last two digits. That is, when 1999 turned to 2000, it was predicted by many scientists that the program would revert to 1900 rather than 2000 and shut down. In response to the persistent media broadcasting, this "likelihood", backed by many scientists, people began hoarding water, gasoline, and survival foods. Some, filled out new wills in anticipation of the Y2K apocalypse. However, when the Time Square Ball in New York City fell, heralding the birth of a new millennium, those computers smoothly transitioned and life continued as normal.

Figure 5-2

Fig 5-2 *The cover of Time magazine in 1999 depicted the end of the world hysteria of Y2K.*

The lessons learned from these examples are once again simple. That is, scientists can be wrong, many of them at times. The media and public can be misled and over respond to a perceived threat. Like the promotion of global warming as a crisis or emergency, the reality revealed over time gives us the true answer. Since "The Inconvenient Truth" was published in 2005, earth is not boiling, droughts and hurricanes are no more common or severe than in the past and the atmospheric average temperature has risen only 0.08 degrees Celsius/decade (0.144-degree Fahrenheit/decade)[14] consistent with solar activity and orbital path variation of the earth around the sun as depicted in Chapter 2.

Example 5: "Those who cannot learn from the past are condemned to repeat it" *George Santayana, Harvard trained Philosopher.*

What began as the Incontinent Truth expanded to a global warming concern, then to a global warming crisis, then to a global warming emergency, then to a global warming apocalypse and now to a boiling earth. This ratcheted up rhetoric and proclamations of a calamity is not science and certainly not proof. The volume of their proclamations and the passion behind it is designed to convince you that it is, but it has nothing to do with the validity of their cause.

Dr. Robert Marx

The most poignant example may only be remembered by those over 50 years of age but is exceedingly noteworthy for those of younger generations to know. It is the "oil shortage" of the 1970s. In the 1976 Democratic Party primaries, all the participants, Jerry Brown, Governor of California, Henry "Scoop" Jackson, Senator from Washington State, Frank Church, Senator from Idaho, Sargent Shriver, US Diplomat, Mo Udall, Congressman from Arizona, George Wallace Governor of Alabama, and then soon to be President, Jimmy Carter all agreed that the world would run out of oil by 2005 or at the latest 2015. Each candidate produced "sound scientific evidence" of the oil running out by then.[20] An oil shortage became an accepted undisputable fact. The oil shortages prompted gasoline rationing. Citizens could fill up their cars only on an odd or even day or an algorithm based on their license plate. Gasoline stations saw lines several blocks long. Highway speed limits were reduced to 55 mph and strictly enforced in an effort to stretch our diminished supply of oil. Gasoline was not even available for many days. (Fig 5-3). Citizens were told to keep their thermostats to

Climate Change: The Hoax of CO₂ Revealed

no higher than 68 degrees Fahrenheit in the winter and to limit their use of air conditioners in the summer. Larger cars were referred to as "gas guzzlers" and were sold only at elevated prices.

Figure 5-3

Fig 5-3 *Picture of a service station taken in the latter half of the 1970s*

These policies caused the deep and prolonged recessions during the Jimmy Carter administration as well as the so-called misery index (inflation &

unemployment)[15] which led to President Carter's failed re-election bid.

History notes that there was no real oil shortage in the 1970s and that all the policies and regulations were unnecessary. Oil reserves existed in the Dakotas, Western Canada, Siberia, The North Atlantic, The Gulf of Mexico, Venezuela and Mexico and the noted oil deposits in the Middle East are not anywhere near depleted as of 2024.

The assuredness of those politicians, media, and scientists as well as the accepted naivety of the public is repeated now regarding global warming. Indeed, "Those who cannot learn from the past are condemned to repeat it."

Climate Change: The Hoax of CO_2 Revealed

References
Chapter 5

1. Gora A. An Inconvenient Truth: The Planetary Emergency of Global Warming. Rodale Books. May 26, 2006
2. Gordon C, Laurie D. The Down to Earth Guide to Global Warming. Google Books 2007
3. Mann ME, Bradley RS, Hughes MK. Global Scale Temperature Climate Forcing over the Past Six Centuries. Nature 1998, 392:779-87
4. Climate Change 2001. Third Assessment Report of the Intergovernmental Panel on Climate Change. IPCC-AR2 2001. Cambridge University Press
5. Douglass DH. Christy JR> A Climatology Conspiracy. American Thinker 9website) December 20, 2009. https://www.americanthinker.comarticles/2009/12/aclimatology conspiracy. html
6. Monford AW. The Hockey Stick Illusion: Climategate and the Corruption of Science.

London: Stacey International.

7. McIntyre S, McKitrick. Corrections to the Mann et al. (1998) Proxy Data Base and Northern Hemispheric Average Temperature Series. Energy & Environment 2003. 14:751-777
8. Mann ME, Bradley RS, Hughes MK. Correction: Corrigendum: Global Scale Temperature Patterns and Climate Forcing over the Past Six Centuries. Nature 2004, 430:105
9. Singer SF. Ot Talk, Cold Science. Independent Institute Publishers, 2021, pp80-84
10. COVID-19 Origins: Experts consulted by Fauci Suddenly Changed their Minds. The Heritage Foundation, Sep. 6, 2023. https://www.heritage.org>publichealth>commentary
11. Wolf Horrell EM. Wilson L. Orazio JA. Melanoma-Epidemiology, Risk Factors, and the Role of Adaptive Pigmentation in Melanoma- Current Clinical Management and Future Therapeutics. Murph M ed., 2015, doi:10.5772158994
12. Guo YY, Zhang JY, Li, XF. PTCH1 Gene Mutations

in Keratocystic Odontogenic Tumors. National Institute of Health 2013.
https://www.ncbi.nlm.nih.gov>articles>PMC3804548

13. Odontogenic Keratocyst (OKC): Reverting Back From Tumor.
Springer Link. https://link.springer.com>article

14. Singer FS. Hot Talk, Cold Science. Independent Institute Publishing
2001 pp153-156

15. Reed E. The Misery Index: Definition and History. Smart Asset May 7, 2023. https://smarlasset.com>financial-advisor>misery-index

Dr. Robert Marx

Chapter 6

The Hypocrisy of the CO_2 Obsessed

Those intent on linking climate change (global warming/sea level rise) to CO_2 have become a virtual cult using proclamations, dire predictions (all of which have not come to pass) and mandates in place of real science. To underscore that their motives are more focused on taxes, regulations and mandates to control US citizens, we need to ask what are they really doing to reduce atmospheric CO_2.

Are They Promoting Reforestation?

The single most effective means of removing CO_2 from the environment is to plant trees. Yet, there is no concerted governmental support or project to accomplish that. If the Biden-Harris administration can appoint former failed presidential candidate John Kerry to a newly developed title "United States Envoy for Climate Change", then it can certainly appoint someone as a "Special Envoy for Reforestation" or at

Climate Change: The Hoax of CO_2 Revealed

least charge the Environmental Protection Agency (EPA) with that worthwhile mission. As I discussed in Chapter 4, the community adjacent to the Mount St. Helens devastation coupled with Weyerhaeuser's commercial support was able to plant 18 million trees in just three years. Why can't our own government support or accomplish similar projects? The reason is lack of leadership and initiative. Perhaps they also know that this book and the nongovernmental International Panel on Climate Change (NIPCC) are proving that man-made CO_2 has nothing to do with the global warming they decry.[1] It is much easier to project a dire environmental fear to the citizenry than to do anything about it except restrict those citizen's freedoms. It is much easier to tax us for not doing what they mandate. It is also much easier to mandate electric cars, solar panels, the elimination of gas stoves etc. than to do anything about CO_2 removal by known means.

Trees remove atmospheric CO_2 by the known chemistry of photosynthesis, i.e. $6\ CO_2 + 6\ H_2O = C_6H_{12}O_6 + 6\ O_2$. The driving force for this biosynthesis is the energy from sunlight. The $C_6H_{12}O_6$ is a sugar molecule (glucose) for plant growth. It is worthy to specifically note that this process has a multiplication factor by taking six molecules of CO_2 out of the atmosphere and sequesters six molecules of water that would otherwise be in our oceans. Once again, the additional benefit of trees is that a single thirty foot tall tree contains 150 to 200 gallons of water in its roots, trunk, branches, and leaves and keeps it out of the oceans.

It seems that our leaders are also unknowledgeable about the fact that seagrass, coral and kelp are three of the most active plants that take up CO_2. If they blame CO_2 as much as they do and lecture us about a "Climate Crisis" and "Climate Emergency" that can only be reversed by taxes and electric cars then one would think they would also be actively pursuing cultivation of aquatic plants and coral the one aquatic animal, to take CO_2 out of our sea water. Why didn't they?

Seagrasses like lawn grass grow by colonial growth (underground root expansion) and/or sexual reproduction (seeds). Their activity can turn over and generate 10 liters of O_2 for each square meter of

Climate Change: The Hoax of CO₂ Revealed

ocean floor and take out 10 liters of dissolved CO_2 from the oceans via the same photosynthesis chemical equation.

Kelp, as exemplified by the giant kelp forests off the west coast of the US, represents the largest seaweed known (Macrocystis). It is also the fastest growing plant on earth[2]. Therefore, it becomes an important biologic tool to remove dissolved CO_2 from the oceans which by the solubility of CO_2 in water will be in turn removed from the atmosphere. The concern over kelp is currently misdirected. It is focused on climate change's effects on kelp (warming ocean) rather than on kelps effect on Climate change.

The biology of kelp is complex. However, we do know that the ideal ocean temperature range for kelp growth is 42 to 72 degrees Fahrenheit[3] which is maintained constant by the Humboldt current.

Above or below this temperature range, kelp growth is suppressed. Essentially, kelp thrives best in cooler waters and explains why it is most prominent on the US west coast where cold water of the Humboldt current flows southward from the Bay of Alaska. This is just the opposite of the ideal water temperatures for coral which is 72 to 84 degrees Fahrenheit[4] and explains why coral is found mostly in the tropical seas about Australia, the Indo Pacific and the Caribbean basin. These range limits offer an opportunity as well as a proof that any increase in ocean temperatures (none has been documented to date) will not affect either seagrass, coral, or kelp. That is, each will expand its territory either northward or southward from its current edges to remain within its ideal temperature range.

However, is there a real threat to seagrass, coral and kelp unrelated to global warming? Yes, it is water clarity. What each of these three ecosystems have in common beyond their ability to take CO_2 out of our oceans and therefore out of our atmosphere is a dependence on sunlight. It is the turbidity of the water in their habitat that disrupts their growth and reproduction most. It comes from our desire to live on or near a coast line where our waste, run off, boating activities among many other activities cloud the water restricting sunlight energy from reaching the sea floor

Climate Change: The Hoax of CO_2 Revealed

where these three ecosystems begin their existence. (See Figure 4-5)

Electricity to Replace Fossil Fuels?

The political-media complex pushes electricity to replace fossil fuels to "save the planet". However, like their global warming predictions and their mandates it is based on false science, ignores real science and also ignores common sense reality. The limitations of electricity as a reliable source of energy are numerous. We can break them down into practical limitations and real scientific limitations.

Practical Limitations of Electricity

The mandates on the public and automakers to convert to electric cars has received significant pushback. The public is just not buying them in the large numbers that were predicted and hoped for by the government. Automobile dealers complain that they have lots full of aging electric vehicles that they cannot sell. Some states are actually removing their original electric car mandates.

Dr. Robert Marx

The reasons for this are mostly well known today. They are:

- Electric cars are high tech and are much more expensive than an equal size gasoline powered car or pickup truck ($5,000 to $25,000 more to purchase)

- Although electric vehicles save on gasoline, oil changes and the fact that electric recharging costs are one half that of a gasoline tank fill up, replacing the EV battery costs $10,000 to $20,000. Also recharging stations are beginning to increase their rates. Recent information disclosed that one overnight home charge of a Tesla adds $8.38 per night and commercial charging station now cost between $8 and $32 per charge.

- Charging stations are few and far between potentially leaving motorists stranded. If you run out of gasoline on the highway, a phone call or road side assistance can provide one to two gallons of gasoline to get you on your way. They cannot as yet come by to recharge your vehicle.

Climate Change: The Hoax of CO_2 Revealed

- The lack of charging stations goes back to the Solyndra scandal. Solyndra was a photovoltaic company that claimed their copper-indium-selenide panels would outperform and revolutionize solar energy capture. In 2009, the Obama Administration loaned the company $535 million to make such panels for public, governmental and private use including the installation of recharging stations nationwide. In 2010, Solyndra filed for bankruptcy. No charging stations were ever made, the $535 million loan of taxpayer money was lost as well as a $25.1 million tax break from the California Alternative Energy and Advanced Transportation Financing Authority.[5] This was an example of either kick back money or at the very least a lack of due diligence when investing in a company of unproven stability.

- Home charging stations cost $1,150 to $2,750 to purchase and install and require a 220-volt outlet that may cost an additional $250 to $800.

- Public charging stations require a minimum of 30 minutes to fully charge an electric vehicle, which is defined as only an 80% charge. This time can be even more depending on the size of the vehicle and therefore the size and capacity of the battery. This time to recharge is only if the recharging station is a level three "Super Charger". A level two charging station requires seven hours and a level one, like home charging stations, requires 12 hours.

- Time considerations: A motorist requiring a recharge may find it necessary to wait in line to access one of the few charging stations before he/she begins his/her 30-minute minimum charge.

- Electric vehicles are costly to insure in general because of the high cost of component parts and the skill it takes to repair them.

- Electric cars and small pickup trucks are certainly clean, quiet and attractive but because global warming is not caused by fossil fuel emissions of CO_2, the owner is not "saving the planet". Certainly, electric vehicles do and will continue to have a place in everyday transportation needs, but it will be limited to smaller, lighter vehicles and shorter trips. This is perhaps underscored by a recent automobile commercial where the driver and passenger are discussing the "paradox" of a plug-in electric car that is also an internal combustion powered car. The automobile company resolves their conundrum by concluding "electric for short trips-gasoline for long trips".

Are Electric Vehicles actually more harmful to the environment than Gasoline-Powered Cars? Most everyone has heard about the fact that much of the electricity used to power electric vehicles (EV) comes from coal. But how much is that and what are the others? It may be shocking to know that 80% comes from fossil fuels (coal 21.9%, natural gas 39.0%, oil/diesel 19%).[6] Nuclear energy produces only 6.9% and renewable sources 13.1%.[6] Therefore, there is a less than advertised environmental benefit for electrical vehicles.

Since electric cars and trucks have been in use, there has been a disturbing finding of excess tire wear coupled with the announced fact that special EV tires are made of much more toxic compounds.[7] It has been found that tires on EV cars wear out two to four times faster and produce more particulate rubber dust (microplastics) due to the weight of their batteries and the excessive torque required to accelerate from a stopped position.[7] The tires particulate pollution is 2,000 times that of a gasoline-powered car's exhaust.[7] This becomes even more disturbing when one considers that 78% of the microparticles found in our oceans is from tire wear.[8] One now needs to be concerned that the push from the climate change obsessed to convert to EV's is not a solution to the environment but a worse environmental risk.

Science Limitations of Electricity

What started the initial climate change advocates on the wrong path was their singular focus on CO_2 from fossil fuel burning. They either ignored the impact of earth's orbital path variations and solar output variations as well as the history of earth's past warmer and cooler periods or purposely withheld them from the public. Now, a similar wrong path course is being undertaken by their belief and reliance on electricity to completely replace all internal combustion vehicles. Electricity simply

Climate Change: The Hoax of CO_2 Revealed

cannot do that. Gasoline and diesel-powered vehicles will continue to be the major transportation mode and industry's use of big machinery. Why?

To answer that, we need to review the basics of electricity from any source. An amp (ampere) is the number of electrons flowing through a wire or circuit. A volt is the force of the electrons flowing through a wire or circuit. An ohm is a unit of electrical resistance. As an example, your oven or toaster has a certain number of electrons running through it measured in amperes. These electrons are pushed by a force measured as voltage. The electrons are impeded by the metal in the toaster or oven, often nickel and chromium alloys, which have a higher melting point. The electrons hit the nickel chromium molecules creating friction and creates the heat of impact which goes on to toast your bread or roast your turkey dinner.

Electricity comes in two forms, Direct Current (DC) and Alternating Current (AC). In direct current, the electrons flow constantly in the same direction. In alternating current, the electrons flow will periodically change their direction.

Dr. Robert Marx

Alternating current is much more powerful than direct current. Direct current comes from batteries and exhibits a greater power loss than alternating current. The much greater power of alternating current and the power loss of direct current is the major limiting factor in electric vehicles which rely on direct current from batteries.

Electric vehicles are limited to batteries made of lithium, cobalt, nickel and graphite.[8] The weight of the battery in an average electric car is 1,000 LBS with some more than 1836 LBS.(Tesla)[10] Therefore, battery weight is the limiting factor in the size of any electric vehicle. The larger the vehicle, the larger the battery needed to power the weight of a larger car or pickup truck. The combined weight of the vehicle, its passengers and the weight of the larger battery becomes the limiting factor to the carrying capacity of the vehicle. This becomes impossible in vehicles bigger than a van, a medium-sized pickup truck, or a small school bus. An example of this may have gone unnoticed by most, but was obvious to those who know the capabilities and limitations of battery power. This occurred in late October of 2023 when

Climate Change: The Hoax of CO_2 Revealed

California Governor Gavin Newsom visited China. Governor Newsome proudly showed a Chinese battery powered bus and made assertions that the US should follow China's environmental leadership and convert to the type of bus he displayed. However, the bus he displayed was no bigger than today's US made vans and had a seating capacity of 10 people. The average US single decker bus has a seating capacity of 40 to 60 people. This represents misrepresentation. The truth of the matter is that such battery powered buses may serve as airport shuttles and local patient buses to a hospital (Jitney's) but not much more.

The limitations of a battery powered vehicle is underscored when we look at the trolley cars in San Francisco, the subways of New York and Chicago and the Electric Trains popular in Europe. After all these are all run on electricity. That is correct, but it is alternating current powered from major land based electric powered power stations. Such systems cannot be powered by batteries. The weight of the battery alone would exceed the hauling capacity of the vehicles. Therefore, for the foreseeable future, most airplanes, trucks from large pickup trucks to 18 wheelers, farm tractors, bulldozers, cranes, and other heavy machinery, among others, will require gasoline or diesel. Why?

The why gets to the very root of the matter, that is born in the natural laws of physics and totally ignored or misunderstood by the climate change obsessed. It has to do with the principle of energy density. The energy density released by burning gasoline is enormous compared to that of electricity. The only energy density greater is from nuclear fission, nuclear fusion or hydrogen. From the comparison of gasoline used in internal combustion engines which has an energy density of 47 MJ/kg (million joules per kilogram)[11] the energy density of a lithium ion-cobalt-nickel- graphite car type battery is only 0.4 MJ/Kg.[12] That means gasoline powered vehicles or machines of any kind create 117.5 times more energy than a battery.

A quote that should be heeded by the Climate Change obsessed.

Climate Change: The Hoax of CO$_2$ Revealed

The Star Trek original episode in 1966, Chief Engineer Scotty to Captain Kirk, "I can't change the laws of physics".

Dr. Robert Marx

References
Chapter 6

1. Idso CD, Singer SF. Climate Change Reconsidered: 2009 Report of the Nongovernmental International Panel on Climate Change (NIPCC). Google Books 2009
2. The Power of Seagrass=medium.https://medium.com>the-power of-seagrass-5ca9774edf...
3. Kelp Forest/Habitat. Monterey Bay Aquarium. https://www.montereybayaquarium.org>habitats>kelp-forest
4. In What Types of Water do Corals Live? - National Ocean Service. National Ocean Service (gov) https://oceanservice.noaa.gov>facts>coralwaters
5. Solyndra Scandal/Full Coverage of Failed Solar Start UP. Washington Post. https:??www.washingtonpost.com>specialsreports>sol...
6. Electric Power Sector Basic/USEPA. United

Climate Change: The Hoax of CO_2 Revealed

States Environmental Protection Agency (.gov) Dec. 18, 2023

https://www.epa.gov>power-sector>electric power-sect...

7. Miznazi A. A rolling Conundrum: Electric cars save gas but tier wear shocks South Florida Drivers Miami Herald Friday, Jan. 26, 2024, Saturday Jan. 27, 2024. Miamiherald.com Column 121, No.134

8. Toxic Tyre Dust: This source of microplastics pollution could be the worst of all. Euronews, Oct. 10, 2023.

    https: //www.euronews.com>Green>Green News.

9. What Materials are Used to Make Electric Vehicle Batteries?

    AZOM, Nov. 2, 2022. https: //www.azon.com>article

10. How much does an electric car battery weight? Hertz Jun. 16, 2023. https: //www.hertz.com>blog>electricvehicles

11. Energy Density. Energy Education. https:..energyeducaiton.ca>encyclopedia>energy_den

12. Schlacter F. The Back Page-American Physical Society. August/September 2012, 21, (8). https://www.aps.org>publications>apsnews>backpage

Climate Change: The Hoax of CO_2 Revealed

Chapter 7

Clearing the Air

Climate change activists realizing that their vilification of CO_2 as the cause of global warming is now looked upon as dubious by the public added a co-conspirator to global warming; Elsie the cow. Yes, in addition to their claims that CO_2 is causing apocalyptic global warming, they have added methane (CH_4) to the post office pictures of their most wanted. Yet, their case against cows is seriously flawed and has as many holes in it as the Swiss cheese that cows produce.

The initial focus of their claim was on cow flatulence. However, to be more comprehensive, they are now including cow burps. Like their animus toward CO_2, their focus on the cow flatulence and burps proceeds with suppositions and projections rather than facts and hard data.

The Case Against Cow Flatulence

Dr. Robert Marx

Cow flatulence has not been accurately measured related to its chemical gas content, its volume nor its frequency. However, human flatulence has and reveals many of the misgiving related to flatulence in general.

Human flatulence occurs approximately 8 to 20 times per day or 3,000 to 7,300 times per year and has an average gas volume of 450 ml to 1400 ml or about 1 pint to 1.5 quarts (0.375 gallons).[1] The chemical composition of flatulence is nitrogen (N_2) 59%, hydrogen (H_2) 21%, carbon dioxide (CO_2) 9%, methane (CH_4) 7%, oxygen (O_2) 4%, and trace amounts of hydrogen sulfide (H_2S), carbon disulfide (CS_2), mercaptans and thiols.[2]

It is these trace amounts of sulfides, mercaptans and thiols that give flatulence its offensive odor. All the other gases, including methane, are odorless and colorless. It should be noted that nitrogen is in the highest concentration, indicating that most flatulence is from swallowed air rather than fermentation in the gut. The oxygen present is also from swallowed air and the CO_2 is from gut cellular metabolism, not too much unlike our lungs.

Climate Change: The Hoax of CO_2 Revealed

Cows are known to take 30,000 to 45,000 chews daily.[3] As in humans, each chew is accompanied by swallowing some air as both species must breathe while chewing; Humans 12-8 times/minutes, cows 30 to 45 times/minute. As in humans, much of their flatulence and their burps are actually mostly air with less methane than thought.

Cows are targeted because they are ruminants and therefore have four chambered stomachs. The chambers in cows are called the rumen, the reticulum, the omasum and the abomasum. It is in the rumen where bacteria induced fermentation takes place and in which methane is produced as a byproduct so that indeed cows produce much more methane than humans (about 15 times more). However, there are over 200 species of ruminants that do the same thing. Sheep, deer, moose, all antelope species, goats, giraffes, yak, musk ox and buffalo are just a few of these species. Pertinent to that is the American Buffalo which are close to cattle in their diet, size and flatulence. Yet, in the year 1500, there were an estimated 60 million buffalo emitting methane.[4] If you recall from earlier chapters, the year 1500 was in the middle of earth's cold period known as the "Little Ice Age".

In addition to ruminants, we must realize that nearly all animals and even insects emit flatulence. Most mammals that are not ruminants, such as humans, also emit flatulence and burps daily. Chimpanzees and all of the ape family do, bears do, elephants do, whales and dolphins do and on and on. So why are the global alarmists focusing in on cows? Simply stated, they are an easy target. Climate alarmists target cows because they can use estimates to replace real numbers and it is an animal known to everyone. Like CO_2, cow flatulence and burps are their means to create acceptance of their flawed accusations and their mandates that allows them to control what you drive, how you heat your home, what you can buy and now what you can or should eat.

The award for the greatest hypocrisy should be given to the United Nations in December of 2023. They conducted their Conference of the Parties (COP 28) meeting in Dubai, United Arab Emirates, one of the richest middle eastern countries due mostly to oil and now tourism.[5] During the meeting, the US envoy for

Climate Change: The Hoax of CO$_2$ Revealed

Climate Change, John Kerry, was suspected of emitting his own flatulence as he spoke about the threat of methane being emitted from cows. The panel members reacted demurely to the odor as Mr. Kerry paused to regain his thoughts and composure. Moreover, the hypocrisy award was cinched when the UN served smoked beef, hamburgers and steaks to most of the 70,000 in attendance. Additionally, most of these seventy thousand attendees flew to Dubai on jets that spewed out tons of CO$_2$ on the way there and on their way back. One must at least give some credit to the COP 28 attendees that the beef they ate produces much less flatulence than a veggie burger.

Much of the methane that exists in the atmosphere does not come from cows but from natural sources such as swamps, decaying sargasso and kelp, and land trash mounds. As previously stated, 200 species of other ruminants and nearly every animal on the planet adds to it.

Is Elimination or a Significant Reduction in the Number of Cows a Good Idea?

Dr. Robert Marx

Since cows are only one source of methane, is it prudent to eliminate them or reduce their numbers? The likely answer is no because their societal worth exceeds their theoretical and overstated contributions to global warming as a single source of methane. It would be prudent to remember that we rely on cows for not only meat protein, but milk, a huge variety of cheeses, cream, yogurt and butter among much more. We also need to understand the many recipes, cakes, donuts, and other pastries that require milk or its products. In addition, jackets, shoes, baseball gloves, hockey gloves, basketballs, and the leather in furniture are derived from cows, as are most capsules in the pills we take and in most all adhesives.

Can Methane Going Into Our Atmosphere be Reduced Without Sacrificing Cows?

The simple answer is yes. The production of methane from cows can be and has been reduced by diet control. Adding small amounts (0.5%) of red kelp to the cattle's feed reduces their methane emissions by 52% as does using oils and grain rather than grass to the diet.[6] One study also claimed an 80% reduction in methane emissions from cow burps by adding red algae to their diet.[7] Private companies, not government, are now gathering cow manure and processing the slurry from it into pure methane known as natural gas which fuels heaters, stoves, ships, and even rockets.[8] Therefore, if we want to, we can use the cow manure for a positive effect rather than an overstated greenhouse gas fear.

Another source of atmospheric methane is leaking pipes. One study estimated 630,000 leaks in the US distribution of current natural gas pipes.[9] However, the government's Environmental Protection Agency estimates 5 times less[10] and neither could identify the actual or even relevant amount of methane released. Suffice it to say that this is a fixable problem. Identify and fix the leaks as you find them. Both the cow production of methane and the methane leaking from aging main line pipes is a fixable problem. Ask yourself, is there governmental support to fix either source and is it more than their support for a carbon tax, carbon credits or that for EV cars?

Does Mother Nature Make the Methane Problem Moot?

Dr. Robert Marx

One of the most known properties of methane is that it is very flammable. This has even been taken advantage of by some usually young men in the form of "lighting farts" and responding with howls of glee and gaining a sense of accomplishment as a blue flame shoots out. After all, methane is natural gas.

The complete earth's atmosphere is 78.09% nitrogen N_2, 20.95% oxygen O_2, 0.93 Argon (Ar), 0.41% carbon dioxide (CO_2), 0.00017% methane (CH_4) and 0.0000007% ozone (O_3). Therefore, methane and ozone are considered "trace gases". Although both methane and carbon dioxide have a controversial estimated longevity in the atmosphere, methane is known to be much less than that for carbon dioxide. This is likely due to lightning. There are 25 to 40 million lightning events in the United States each year, 1 to 1.4 billion worldwide.[11,12] The average temperature of lighting is 53,500 degrees Fahrenheit, which is five times hotter than the surface temperature of the sun and well above the 1,076 degrees Fahrenheit ignition temperature of methane. Simply stated, methane by whatever source does not have an overtly warming effect on the earth due to

Climate Change: The Hoax of CO₂ Revealed

mother nature's tool of lightning. The public can feel free to emit flatulence, wear leather, drink milk, and eat a hamburger without feeling guilty.

Dr. Robert Marx

References
Chapter 7

1. Everybody Farts. But here are 9 surprising facts about flatulence you may not know. Vox, Aug. 11, 2015.
   https://www.vox.com>2014/ 12>fart.flatulence
2. What is the chemical Composition of Gases During Fart? Quora. May 8, 2015.
   hltps://www.quora.com. What is the chemical composition of gases during a fart.
3. Amaral Phillips DM. Why do Cattle Chew Their Cud? - animal and food sciences University of Kentucky.
   https://afs.ca.uky.edu>content>whydocattlechewtheir cud?
4. American Bison/Maymont Foundation.
   https://maymont.org> ...animals>wildlifehabilats
5. UN Climate Change conference-United Arab Emirates. Jan. 12, 2024. United Nations

Framework Convention on Climate Change. https://unfccc.int>cop28

6. Van Deelen G. Feeding cows Seaweed Reduces Their Methane Emissions, but California Farms are a Long Way From Scaling up the Practice. Inside Climate News. Jun. 14, 2022. https://insideclimatenews.org>news

7. Study Finds that Red Seaweed Dramatically Reduces the Amount of Methane that Cows Emit, with Emissions from Cow Belches Decreasing by 80% r / science

8. The Power of Cow Manure: How farms of the future are transforming waste-to-energy operations. POWER Magazine. https://www.powermag.com>the-power-of-cow-man...

9. Weller ZD, Hamburg SP, Von Fischer JC. A Natural Estimates of Methane Leakage from Pipeline Mains in Natural Gas Local Distribution Systems. Environ. Sci Technol. 2020, 54, 14, 8958- 8967

Methane Emissions from US Gas Pipelines Leaks. Environmental Defense fund.

https://www.edf.org?default>files>documents (PDF). August 2023

10. Lightning Strike Victim Data/Lightning/CDC. Centers for Disease Control and Prevention (.gov) https://www.cdc.gov>disasters>lightning>victimdata

11. 10 Striking Facts about Lightning. Met Office https://www.metoffice.gov.uk>thunder-and-lightning

12. Understanding Lightning: Thunder. National Weather Service (.gov) h tLps: // www.weather.gov>safety> lightning-science-thunder

13. The auto-ignition temperature of methane/US EPA United States Environmental Prevention Agency (.gov) https://hero.epa.gov>referene>details>reference-id

Climate Change: The Hoax of CO_2 Revealed

Chapter 8

A Warm Fuzzy Feeling

One of the focal points of this book is to describe the mechanism of a gradually warming earth consistent with the recorded warming periods of the Roman Warming Period and the Medieval Warming Period.[1] As indicated in Chapter 2, the mechanism for these warming periods are the centuries long variations in solar output and the longer-term variations in the earth's orbital path around the sun.[2]

Another focal point of the book has been to debunk the hysteria of a rapid global warming from CO_2 and methane. The assertions made by the climate change obsessed that global warming is apocalyptic are denied by the simple facts of chemistry and physics and are supported by the fact that all of their predictions and computer models have been wrong.

We all need to realize that earth is emerging from the last ice age as it did from each of the other four. All past ice ages ended as a result of natural cycles irrespective of man's production of CO_2 or cows' production of CH_4. Each past ice age was followed by a long-term warm climate that spawned the age of several prehistoric species before, during and after the dinosaurs. During these interglacial warming periods, there were also cooler and even warmer periods within it that were several centuries long, without any influence from humans. Earth's history is telling us we are entering a slow long-term warming trend that cannot be reversed or significantly altered by avoiding fossil fuels or by a partial or complete conversion to solar or electrical power.

Is a Warmer Climate a Good Thing or a Bad Thing?

History notes that warmer climates have been beneficial to our species and many other species as well. This can be broken down into three time frames: the transition from the last ice age, the Roman Warming Period, and the Medieval Warming Period.

The Transition from the Last Ice Age:

Climate Change: The Hoax of CO_2 Revealed

There is no concrete evidence that tribes of human beings inhabited North and South America before 15,000 years ago.[3] As the last ice age was waning, the Bering Strait land bridge allowed early human immigrants from Northeastern Asia to migrate to the Northwestern base of Alaska. These were small bands still living under harsh conditions. The cold and remaining glacier slowed down and inhibited further travel, suppressed their birth rate, enhanced the infant death rate, and kept the longevity of adults from increasing. Around 13,000 years ago, the planet sufficiently warmed to slightly mitigate these harsh factors. This resulted in the findings of the earliest humans known in the United States as the "Clovis People" 13,000 years ago.[3] This site is named after its location in Clovis New Mexico. Of course, other tribes migrated, as did the Clovis people. The Clovis people certainly had to travel the 3,500 miles by foot to reach New Mexico. Others certainly migrated south from the Alaska Bering Straits area as did the Clovis people. Their remains and artifacts have yet to be discovered. Obviously, the migration began hundreds or even a thousand years earlier.

Dr. Robert Marx

As the climate continued its post ice age warming, food supplies in the form of plants and game became numerous and more available. Even though the harshness of the winters and a rugged existence still existed, it was less than that of the last ice age. With animal skins more obtainable and a readily available food supply like the herds of American Bison and other game, the birth rate rose and the death rate declined. These initial small bands grew into tribes and expanded throughout North America and on into Central and South America.

Each tribe expanded its territory as its population grew. By 1500 there were 565 separate Native American tribes[4] for a total population of 1.8 million in North America.[5] In Central and South America called Meso-America several cultures arose: The Olmec's 1600 to 400 BCE,[6] The Mayans 250 BCE to 900 AD with 40 cities and an estimated population of 7 million[7] and the Aztecs 1300 to 1521 AD when the Spanish Conquistador Hernan Cortez vanquished them.[8]

Climate Change: The Hoax of CO_2 Revealed

One can obviously see the explosion of life and culture that took place in roughly the past 15,000 years of a warming, moderate climate.

The Roman Warming Period

The Roman Warming Period 200 BCE to 400 AD followed a significant cooling period. Prior to the Roman Warming Period the Egyptian, Greek, and Roman cultures developed a foothold mostly due to their location in the warmer middle eastern belt of the Mediterranean. The Germanic tribes further north were still mostly scattered tribes and the Vikings had yet to emerge. When the Roman Warming Period was in full force, these three cultures expanded and reached their apex. This was largely due to longer growing seasons, better crop yields and fewer crop losses due to freezing. This once again increased individual longevity, increased birth rates, decreased death rates and supported the development of their arts. It was not until the next warming period that the Vikings arose to notoriety and European cities grew and prospered.

The Medieval Warming Period

Dr. Robert Marx

Following the "Roman Warming Period" the "Dark Ages Cooling Periods" began. Of course, history notes the collapse of the Egyptians, Greek, and Roman empires due to wars, invasion of the Mongolian Hordes, internal strife and decadence. The cooler climate also saw a retraction of their crop yields, shorter growing periods, and crop losses from colder weather and frost contributing to the decline of each empire.

The medieval warming period from 900 AD to 1400 AD saw an expansion of European culture and population. What were once tribes became cities by 1200 AD i.e. Paris 160,000 population, Constantinople 189,000 population and Palermo which had declined to 350 citizens during the Dark Ages cooling period rebounded to 135,000 population by 1200 AD.[9] In particular the far north experienced sufficient warming to open the ice blocked Fjords to the Viking ships for their exploration and raiding conquests. A milder climate allowed the Vikings to increase their population and begin their well-known travels, conquering, and their development of settlements in previously unsettled areas. Of

Climate Change: The Hoax of CO_2 Revealed

particular note are the recently discovered Greenland settlements 1100 to 1200 AD some of which are under a glacier today due to the "Little Ice Age" of 1400 AD to 1850[10] AD. Taking advantage of the medieval warmer climate and the more navigable waters they also developed early settlements in North Eastern Canada and the Islands off Newfoundland by 1400 AD.[11] But alas, nature reneged on the Vikings with the return of a very cold "Little Ice Age", which began around 1400 AD that saw these early settlements abandoned and the Fjords once again

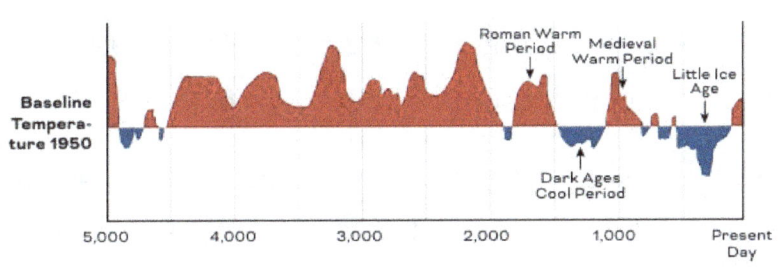

choked with ice (Fig. 8-1).

Figure 8-1

Fig.8-1 *Several societies prospered and thrived during both the Roman warming period and the Medieval warming period.*

Dr. Robert Marx

Is a Warmer Climate Still Beneficial for Humans in the 21st Century?

The answer to this question is certainly yes. Available data confirm this.

Over the planet's history, the benefits of a warmer climate have been mostly that of agricultural productivity and reliability, translating into increased survival, increased birth rates and decreased death rates. More recently, our current warming trend has resulted in a reduction in the undernourished as published by the Food and Agriculture Organization of the United Nations (FAD). The UN reported 216 million less undernourished globally from 1990 to the present. This represents a reduction of 21.4%.[12]

Actually, rising CO_2 levels can be looked at as a blessing. That is, population growth has stressed nearly every country, causing localized famines, food shortages, the highest rate of immigration ever, and numerous regional conflicts among other stressors. The mere fact that CO_2 is the basic nutrient for nearly all earth's plant life, its rise can be expected to

Climate Change: The Hoax of CO_2 Revealed

increase the land that can support agriculture, grow larger crops, and grow more crops. This is exampled from a study that showed a given plant community that absorbed 54.95 pentagrams (Pg) of carbon in the form of CO_2 per year in 1961 increased to 66.75 PG carbon in the form of CO_2 per year in 2010, commensurate with the slow rise in atmospheric CO_2 (note one pentagram of carbon equals 10^{15} or 1 quadrillion grams of carbon or 220 billion pounds of carbon).[13] Such numbers identify the enormous uptake of CO_2 from the atmosphere by plants as well as the reliance of agriculture on CO_2.

While some have predicted worldwide starvation and famine, the world's farmers to date have increased their food production at a faster rate than population growth (Fig. 8-2).[14] Food shortages and regional famines are caused by supply chain and distribution failures not food production.

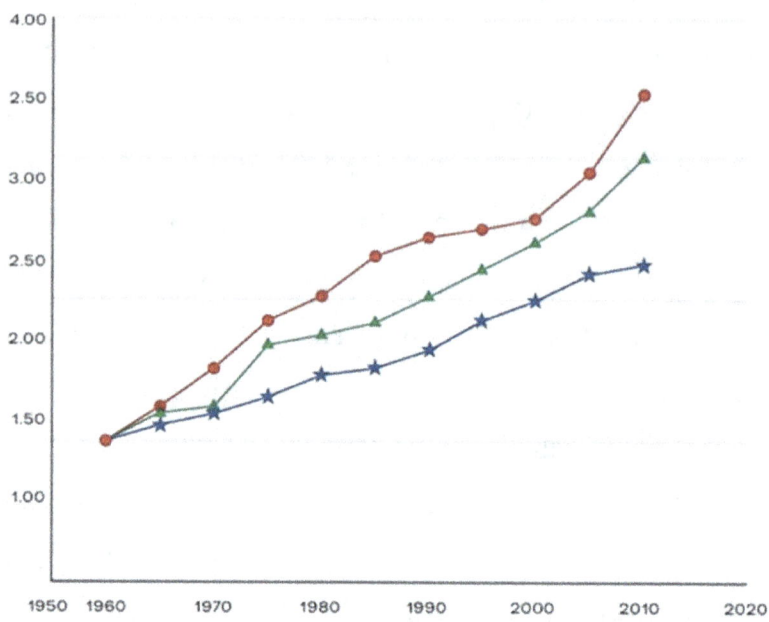

Figure 8-2
● CO₂ emmissions
▲ Food production
★ Population
Source: Idso, 2013, 24 Figure 8

Fig. 8-2 *Graph of population growth, food production and CO_2 emissions normalized to equal values of 1960 plotted to 2010 shows food production increases parallel to CO_2 emission increases and food production increases both pacing ahead of population growth.*

Gradual Global Warming's Support of Human Health

The COVID-19 pandemic focused much of the population's attention on viral disease and disease transmission. While some bacteria and most fungal diseases are promoted by a warm moist climate, most viruses prefer colder drier climates. The classic flu season is not midsummer, it is fall and winter. This is seen every year and why "flu shots" are recommended in October and November in the Northern Hemisphere. This goes for SARS, COVID, RSV, Bird flu, Swine flu and most all others seen in the recent past. A study by Gasparrini et al in 2015 before the COVID-19 pandemic emphasized this related to environmental temperatures of: extreme cold, moderate cold, moderate heat and extreme heat. (Fig. 8-3).[15] One can see across every country extreme cold and moderate cold led the way in deaths attributable to weather. Perhaps the following quote from an accomplished Climatologist and pioneer climate researcher sums it up best.

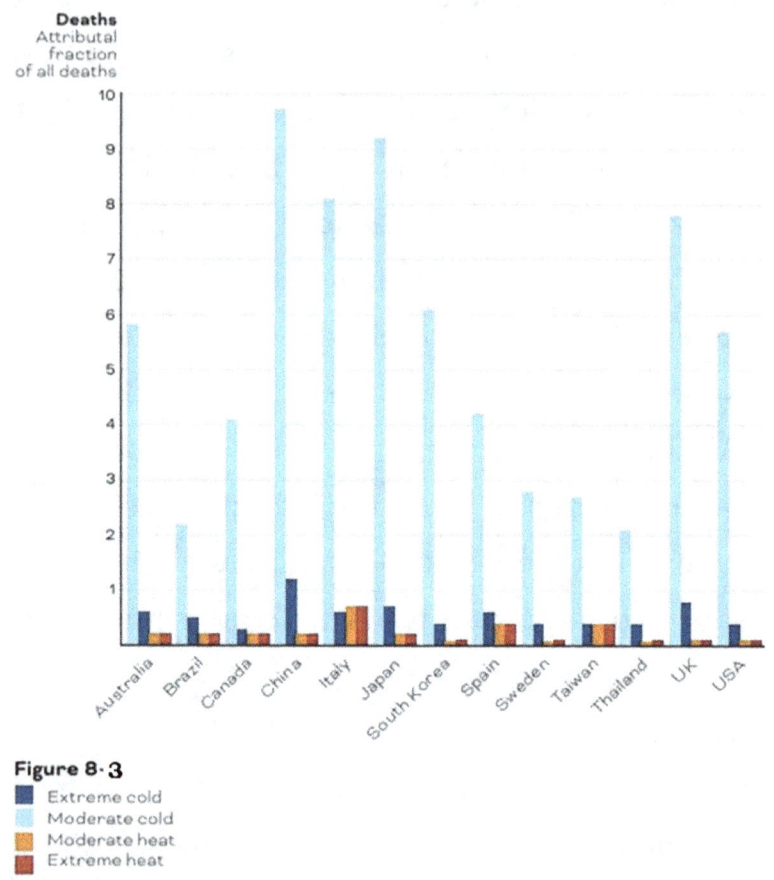

Figure 8-3
- Extreme cold
- Moderate cold
- Moderate heat
- Extreme heat

Source: Gasparrini et al 2015, 369

Fig. 8-3 *Deaths caused by heat versus cold temperatures graphed for 13 industrialized countries. Note: moderate cold temperature caused far more deaths than even severe cold or severe heat.*

"If the world is likely to see only 0.6 C (1.08 F) warming in the coming century then there is certainly no reason to tax or ration fossil fuels, subsidize

Climate Change: The Hoax of CO_2 Revealed

alternative energies such as wind and solar power or engage in the many other costly policies advocated by environmental activists and the policymakers who pay heed to them. People living in a modestly warmer world will be more prosperous, better fed and healthier than people alive today and they will enjoy a lusher and more bountiful natural world as well. What a pity it is that so few people know that this is the Real Story about Climate Change."[16]

S. Red Singer

BEE, PhD

Dr. Robert Marx

References
Chapter 8

1. Alley RB. The younger Dryas Cold Interval as Viewed from Central Greenland. Quaternary Science Reviews 2000,19;213-226.
2. Rayno ME, Huybers P. Unlocking the Mysteries of the Ice Ages.
   Nature 2008, 41, 254-285
3. The First Americans. National Endowment for The Humanities (.gov) https://www.neh.gov>humanities>marchapril>feature
4. List of Native American Tribes from 1500 to 1600. Sciencing, Nov. 22, 2019. https://sciencing.com>scienceprojects>nature
5. North American Indian Population Size AD 1500 to 1985. Wiley Online library. https://onlinelibrary.wiley.com>doi>pdf>ajp ...
6. Olmec Civilization. National Geographic Society. https://www.nationalgeographic.org>encyclopedia. Oct. 19, 2023
7. Maya Civilization. Wikipedia.

https://en.wik.ipedia.org?wik.i>maya civilizataion

8. Aztec Civilization - National Geographic Education Oct. 19, 2023 National Geographic Society.
hllps://eductiona.nationalgeographic.org>resource

9. Largest Cities in Western Europe 1200/statista. https://www.statista.com>society>historical data. Mar. 1, 1992

10. Zwally HJ. Growth of Greenland Ice Sheet: Interpretation. Science 22 Dec. 1989, 246, 4937 pp 1589-1591.
Doi:10.1126/scoemce.246.4937.1589

11. The Norse in Newfoundland/ Adventure Canada. Adventure Canada. https://www.adventurecancada.com>the-norse-in newfoundland...

12. FOA (Food and Agricultural Organization of the United Nations). 2015. The State of Food Insecurity in the World 2015. Rome-Food and Agricultural Organization of the United Nations

13. Zhu Z, Pia Z, Zhu S, et al. Greening of the Earth

and Its Drivers. Nature Climate Change 6:791-795

14. Idso CD. The Positive Externalities of Carbon Dioxide: Estimating the Monetary Benefits of Rising Atmospheric CO_2 Concentrations on Global Food Production. Tempe AZ Center for the Study of Carbon Dioxide and Global Change

15. Gasparini A, Guo Y, Hashizume M. et al. Mortality Risk Attributable to High and Low Ambient Temperature: A Multicounty Observational Study Lancet. The Lancet 2015, 386:369

16. Singer SF. Hot Talk Cold Science. Independent Publishing 2021, p 160.

Climate Change: The Hoax of CO_2 Revealed

Chapter 9

The Real Existential Threats

The climate change obsessed tried to convince the world's citizens that CO_2 is the villain, producing rapid global warming and sea level rise as an "existential threat." Yet, every one of their apocalyptic predictions has been wrong, the knowledge of earth's temperature history has been ignored, and their knowledge of basic chemistry, physics and biology is seriously flawed. Their obsession to turn a natural and gradual rebound from the last ice age as well as year to year and decade to decade temperature variations into a profound fear is shameful. Perhaps, their hoax itself is the real existential threat.

On September 11, 2023, President Biden claimed that climate change was more of an existential threat than nuclear war.[1] The specific quote was "The only existential threat humanity faces more frightening than nuclear war is global warming." Representatives of the climate change obsessed made up false figures. That is, his statement went on to state "The global temperature is rising more than 1.5 degrees Celsius (2.7 degrees Fahrenheit) in the next two decades is scarier than nuclear war." The fact is that the current rate of global warming is at 0.6 degrees Celsius (1.08 degrees Fahrenheit) over the next century, not over the next two decades.[2]

Using more vetted data and the reality observed by everyone, there are more realistic threats that the climate change obsessed have ignored and from which they have distracted the public's attention.

These are the More Real Existential Threats:

1. **Overpopulation**: The world's population exceeded 8 billion in February of 2023.[3] Figure 9-1 illustrates the stark reality of the ascending curve of population growth in less than 75 years (1950 to 2023). It is more than doubling every 40 years. This ascendance is actually an understatement as many individuals are not counted due to inaccurate census taking, avoidance due to criminal activity, gangs, enslavements that are

Climate Change: The Hoax of CO_2 Revealed

well recognized as occurring in several countries in Africa and in China. There are also countries that do not participate in census counts, etc. Even at 8 billion people, one should note that normal respiratory physiology indicates that this number of individuals will exhale 3 billion tons of CO_2 into the atmosphere each year and this amount will continue to climb as the population increases further. Each individual added will need housing, food, transportation, etc. Each will create bodily and other waste, compete for resources, require building material, need health care, require education and training, and will use energy most likely based on fossil fuels despite all the empty claims of the climate change obsessed.

Dr. Robert Marx

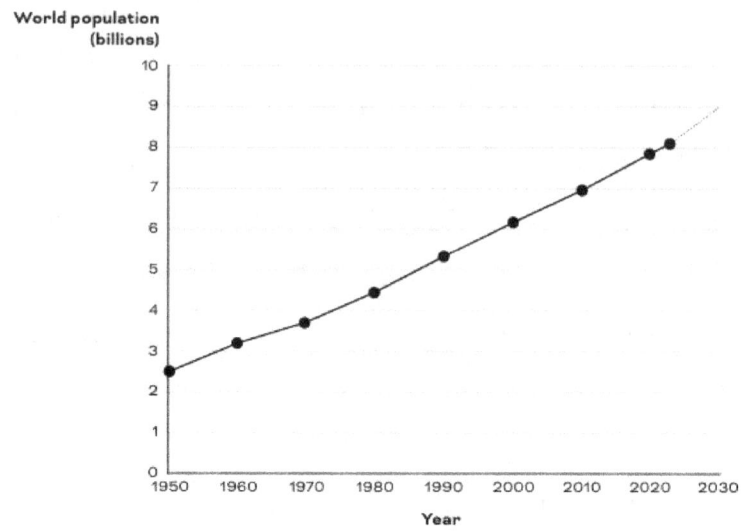

Figure 9-1

Fig 9-1 *World's population growth over time indicates the doubling of earth's population ever 40 years and the likelihood it will double sooner than in the next 40 years.*

In 2022, the population increased by 143 million, which pushed the 2023 population over the 8 billion mark. Each individual will add to the stress of overpopulation as seen today in local conflicts and major wars, mass migrations, and disease outbreaks. Overpopulation is similar to a chain reaction as each individual will need a multiplicity of tangible supports and will affect the environment in a multiplicity of ways. The underpinning of overpopulation is that it is a true existential threat that is insidious, progressive, and persistent.

Local Conflicts and Wars: A byproduct of overpopulation is war. One only needs to look at the world today to confirm this. The war in the Ukraine and the war in Israel represent major conflicts. However, there were 30 other active smaller scale conflicts going on in 2023, such as insurgencies in the Sudan and Morocco, the takeover of the entire country of Haiti by gangs, as well as in several countries in Africa and Indonesia. The root cause of most of these is a conflict over territory, as exemplified by the Gaza Strip, the Crimea and the Donbas region of the Ukraine. Territorial expansion is needed to accommodate each country's growing population and the demands and needs of their citizens.

Mass Immigrations: The current United States Southern Border Crisis represents the most well-known mass migration with over 8 million illegal immigrants into the United States from 2021 to 2024. However, Argentina, the Netherlands,

Dr. Robert Marx

Switzerland, Japan, and Singapore also have serious immigration issues, some over 2 million immigrants from 2000 to 2024. Even Germany (second to the United States) and Saudi-Arabia (third to the United States) have seen mass immigration numbers.[4]

The immigrants are migrating due to repressive regimes, poverty, governmental corruption whereby their leaders ignore the needs of their citizens, and limited available resources in their country of origin. The increased numbers in their country of origin have out stripped their countries infrastructure and resources to support them as well as the will of their leaders to address their needs. Even in the United States, the influx of migrants has resulted in an inability of even the declared sanctuary cities to house them, feed them, remove their trash, cope with increased human excrement, drug inflow, provide health care services and more. The fact that so many countries are experiencing such immigration problems underscores that overpopulation is indeed a fundamental existential threat.

2. **Toxic Pollution**: The second real existential threat is toxic pollution. Industrial waste and careless industrial companies are poisoning our waterways and cities. Would anyone reading this book dare to drink out of the Ganges River, The Lower Mississippi River, Ohio river, Chicago River, Hudson River, or the Blue Danube, which is green due to its pollution? These are just obvious examples for all of us to see. Each reader of this book will know of closer to home examples.

The famous red tide outbreaks in the Gulf of Mexico and around the shores of Florida are caused by runoff from overly fertilized crops.[5] The high phosphate concentration of these fertilizers promotes the overgrowth of over 50 species of red algae, the most common of which is Karenina brevis.[6] These algae produce a potent neurotoxin known as brevetoxins that kills fish and turns the water a dusky red with the odor of decaying fish teeming with bacteria. A similar but less common form of red tide occurs off the California Coast due to several species of algae known as Heterosigma akashiwo, pseudo-nitzschia or Cyanobacteria also known as blue-green algae[7]

Closer to home, in the inland area of the United States, examples such as the Camp Lejeune water contaminations are numerous. As early as the 1980s, two water supply systems at that marine base were found to contain trichloroethylene (TCE) and perchloroethylene (PCE).[8] Both of these are carcinogens that came from industrial wastes. Today, a host of cancers have been linked to this contaminated water supply. Cancers such as leukemia, breast cancer, lung cancer, kidney cancer etc. and additionally infertility and birth deformities.[9]

Another stark reality is that the Ohio River system has been labeled by the Environmental Protection Agency as "one of the most toxic watersheds in the country".[10] At times, after heavy rainfalls, sections

Climate Change: The Hoax of CO_2 Revealed

of the river have been declared unsafe for even recreational use by the Ohio River Valley Water Sanitation Commission. Added to this is the well-known East Palestine train derailment which contaminated acres of ground as well as released tons of butyl acrylate into the nearby creeks and rivers that flowed into the Ohio River just 16 miles away.

Examples such as these are endless. Suffice it to say, toxic pollution has been linked to cancers, fetal deformities, low sperm counts, female infertility and numerous chronic illnesses. Somewhat linked to overpopulation, toxic pollution is an ongoing and increasing global existential threat.

3. **Disease Outbreaks and Epidemics**: The famous and prophetic quote of Plato repeated by General Douglas MacArthur in his farewell address to the cadets at West Point in 1962, "Only the dead have seen the end of war"[11] can be paraphrased to apply to plagues and pandemics today. Perhaps

we should all heed the notion that "only the dead have seen the last of plagues".

Of course, we must start with history's worst plague, the Black Death 1347 to 1353 that resulted between 50 and 80 million deaths and killed 50% of the total population of Europe during those six years.[12] The Black Death was caused by a bacterium (Yersinia Pestis) readily susceptible to today's antibiotics.[13] However, the immune naïve and stricken individuals without the antibiotics we rely on today resulted in an inability of most to survive the disease. The Black Death can be traced back to Marco Polo's opening of the trade routes from China through the middle east and into Europe known as the "Silk Road". This created a transmission vector whereby the fleas infested with Yersinia Pestis living in the fur of the oriental rats gained access to a new territory.[14] The rats proliferated in the trash and discards of Europe's expanding urban cities, allowing their fleas access to pets and people, injecting the bacteria with each bite. The lesson from this early and most lethal plague is that it requires an immune naïve population, a transmission vector and an absence of infection prevention and treatment schemes.

From that time forward, there were five bacterial pandemics involving mostly cholera and sporadic resurgences of the Black Plaque between 1894 - 1940 totaling 15 million deaths. However, since then the most serious pandemics have been viral diseases. This is emphasized by the fact that once Columbus landed in the new world, the population of Native Americans was around 54 million but was reduced to just 6 million by 1900, a loss of 90% of their total population.[15] This was mostly due to smallpox a viral disease to which the Native Americans were immune naïve. Their interface with the "white man" was the vector of transmission and although vaccines made smallpox extinct as of 1980[16] there was no treatment or vaccines available at that time.

Dr. Robert Marx

Since then, the Asian flu claimed two million lives in 1957 and the Hong Kong flu another two million in 1968. However, these pale in comparison to the Spanish flu from 1918 to 1920 and the more recent HIV and COVID pandemics. (Table9-1), (Fig 9-2)

Table 9-1

Disease	Years	Deaths
Black Plague	1347-1355	50-80 Million
Native American Post Columbus diseases	1492-1900	48 Million
Cholera	1817-1827	5 Million
Cholera	1829-1887	18 Million
Russian Flu	1889	4 Million
Black Plague Resurgence	1894-1940	15 Million
Spanish Flu	1918-1920	50-70 Million
Asian Flu	1957	2 Million
Hong Kong Flu	1968	2 Million
Scattered Ebola Outbreaks	1976-Present	13 Million
HIV-AIDS	1980-Present	33 Million
Covid	2020-Present	27 Million

Table 9-1 *Deaths from notable plague/pandemics associated with the years over which they occurred.*

Climate Change: The Hoax of CO_2 Revealed

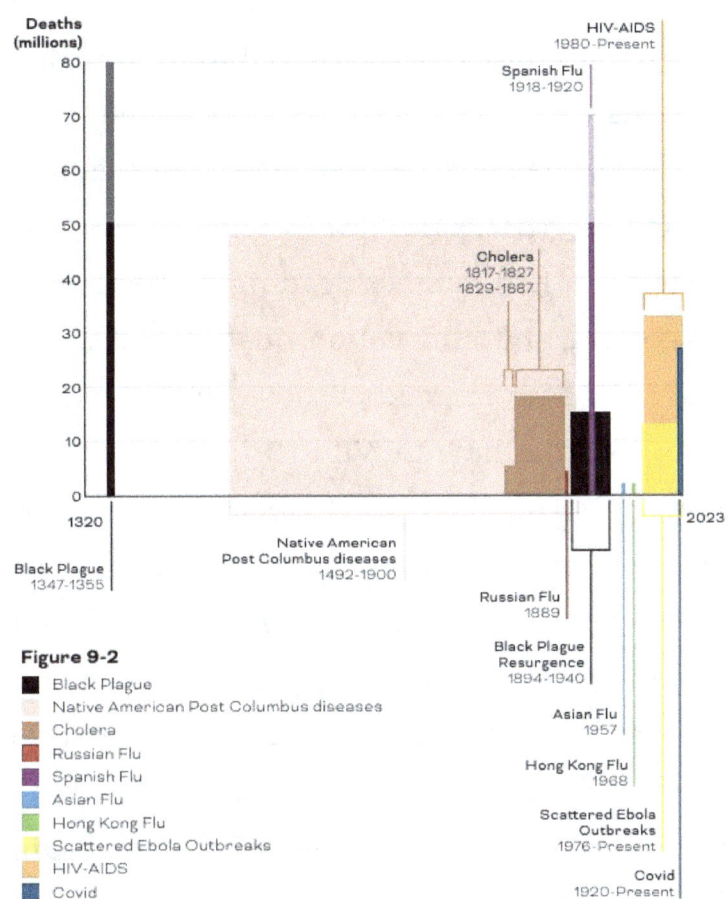

Fig 9-2 *Graph of deaths caused by notable plagues/pandemics through recorded history.*

The Spanish flu actually originated in Kansas within the United States and was caused by an H1N1 influenza A virus.[17] If you're curious, H, refers to the virus hemagglutinin subtype of which there are 18. The N1 refers to the neuraminidase subtype of which there are 11. These are both proteins found on the outer wall of the virus and are associated with transmission, infectivity, and virulence. The H1N1 virus caused a "cytokine storm" in its victims, leading to extreme lung congestion and suffocation deaths that mimic the cause of death in most of those who succumbed to COVID in the COVID-19 pandemic 2020–2022.[18,19] The Spanish flu claimed an estimated 17 to 50 million lives and disappeared not by vaccines or antimicrobials but by the natural means of viral evolution. That is, mutation to a less virulent form due to inter viral competition for susceptible hosts finally selecting out less virulent strains that did not kill the host. Once again, the Spanish flu met all the requirements for a pandemic; a susceptible immune naïve population (the malnourished and crowded populous after World War I), a person-to-person transmission once again due to post World War I encampments and no available antiviral therapeutics or vaccines to stave off its spread.[17]

Climate Change: The Hoax of CO_2 Revealed

The HIV pandemic emerged in 1978 and spread quickly throughout IV drug users, the men having sex with men culture and the blood bank system, particularly one large scale blood bank in Haiti that contaminated both the blood donor and the blood recipient resulting in falsely accusing and ostracizing Haitian citizens.[23] The HIV/AIDS pandemic resulted in 33 to 51.3 million lives lost from 1978 to 2022.[21] The blood banks throughout the world developed procedures and standards that resulted in safe blood transfusion by 1985. By 2000, various combinations of four potent antiviral drugs initiated the decline of HIV and AIDS and its related deaths. Vaccines that were developed did not work because HIV, being a retrovirus, integrates itself into the host's genome so that the host's immune cells were unable to recognize it as a foreign pathogen.[22]

The HIV/AIDS experience met the requirements to become a pandemic but was assisted by the government intervention for some good and some not so good. Once again, the HIV virus met an immune naïve population. It took advantage of three directly injectable transmission vectors e.g.: IV drug use, men having sex with men, and an unprepared blood banking system with no universally accepted safety standards at the time. Since vaccines were unable to be used, the United States and French scientists came to the forefront, developing highly potent antiviral drugs which saved millions of lives and are continuing to do so. The governmental support of funding the antiviral therapeutic development was the good. However, its lack of clarity about its true transmission vectors, prevention methods as well as its inability to effectively address IV drug addiction was the not so good.

The COVID-19 outbreak began in December of 2019 and rapidly gained worldwide attention by February of 2020. Current evidence is conclusive that the COVID virus emanated from the Wuhan lab in China due to their lack of maintaining a level 4 safety protocol.[23] The uniqueness of the

Climate Change: The Hoax of CO$_2$ Revealed

COVID-19 pandemic is that all governments stepped in to mitigate the spread and death rates. However, they and the World Health Organization made numerous errors in its management, prevention schemes, its overreliance on hastily manufactured vaccines, shutdowns, and a fundamental failure to recognize the value of therapeutics leading to confusion and worsening COVID-19's impact on the population. COVID-19 is estimated to have caused 27 million deaths worldwide.[24] Once again, the requirements of a pandemic were met by the COVID virus. Launched onto the unaware and immune naïve public, it spread by the transmission vector of modern day international and regional travel. The prevention schemes of masks did not work well, and the vaccines proved that the COVID virus could easily mutate into a resistant form. This was proven by the vast members of doubly vaccinated and boosted individuals who still developed COVID as exemplified by both President Trump and President Biden. The COVID experience underscores that even with modern medical technology and government mandates, a single virus can create a multi death pandemic.

This review of an incomplete list of multiple death pandemics illustrates that pandemics are getting worse, getting more numerous, more deadly, and that even modern medicine and societal attempts to restrict and mitigate its effect on the population have largely failed. COVID is now waning for the same reason that the Spanish flu became extinct. That is, the evolution of any pathogenic virus to a less virulent form that infects but that does not debilitate or kill the host. The less pathogenic strain can then be transmitted over a longer period of time compared to the more virulent forms that debilitate or kill the host before the host comes into contact with and transmits the disease to other hosts.

Indeed, future pandemics will represent an existential threat boosted by the commonality of national and international travel, governmental interference with a comprehensive medical

Climate Change: The Hoax of CO_2 Revealed

approach, ill thought through mandates, and inept leadership.

If we compare overpopulations, toxic pollution and future pandemics to the hoax of CO_2 and the hyperbole of the climate change crisis, we can see which are the real "existential threats".

Dr. Robert Marx

References
Chapter 9

1. Biden Says Climate Change Poses Greater Threat Than Nuclear War. Bloomberg. https://www.bloomberg.com>news>articles.biden-sa...

2. Singer SF. Hot Talk Cold Science. Independent Publishing 2021. P160

3. World Population Clock:8.1 Billion People (live 2024). World Odometer. https://www.worldodometers.info>world population

4. 10 Countries that take the Most Migrants. US News & World Report. https://news.wgcu.org>news> UF-study-human-action...

5. UF Study: Human Activity Provides for Longer, Stronger, Red Tides. Bayles T. WGCU. https://news.wgcu.org>news>UF-study human-actvi...

6. About Red Tides in Florida. Florida Fish and

Wildlife Commission FWC. https://myfwc.com>research>redtide>general>ab...

7. Toxic Algae Bloom Suspected in Dolphin and Sealion Deaths in Southern California. June 16, 2023. NOAA Fisheries (.gov). https://www.fisheries.noaa.gov>feature-story-toxic-...

8. Camp Lejeune, North Carolina/ ATSDR-Background. Agency for Toxic Substances and Disease Registry (.gov). https://www.atsdr.cdc.gov>sites>background

9. Camp Lejeune Water Contamination Tied to a Range of Cancers, CDC Study Says Jan. 31, 2024. US News & World Report. https://www.usnews.com>news>beststates

10. Cory C. Environmental Protection Agency calls Ohio River the most Polluted in Country. The News Record.
https://www.newsrecord.org>news>environmental - pro...

11. General Douglas MacArthur/Speech to West Point Cadets 1962. MacArthurmilwaukeeforum.com.

https://macarthurmilwaukeeforum.com>resources>mac...

12. The bright side of the Black Death-American Scientist.
hllps://www.americanscientist.org>article>thebright-si...

13. Black Death/Definition, Cause Symptoms, Effects, Death Toll Britannia. https//www.britannia.com>...Accidents & Disasters

14. The Black Death/Western Civilization- Lumen Learning Lumen Learning. https://courses.lumenlearning.com>chapter>theblack-

15. Woodward A. European colonizers killed so many indigenous Americans that the planet could down, a group of researchers concluded. Science Feb. 9, 2019. Business Insider.
https://www.businessinsider.com>news

16. Smallpox/CDC. Centers for Disease Central and Prevention (.gov). https://www.cdc.gov>smallpox

17. Spanish Flu: What is it, causes, symptoms & pandemic Cleveland Clinic. https//myclevelandclinic.org>health>diseases>2177

18. Purple Death: The Great Flu of 1918-

PAHO/WHO. Pan American Health Organization https: / /www.paho.org>who-we-are>history paho>pur...

19. Oud L, Garza J. The Contribution of COVID-19 to Acute Respiratory Distress Syndrome Related Mortality in the United States. J Clin Med Res. 2023, May 31. 15(5):29-281

20. Quammen D. Spillover. W.W. Norton Publ.l New York 2012 99484- 487

21. Global HIV & AIDS statistics-fact-sheets

22. Venkatesen P. HIV Vaccine Trial Failure. The Lancet Infectious Diseases. Apr. 2023:23(4)p410. Published April 2023.
https: //doi.org/ 10.1016/ S1473-3099(23)00148-2

23. Rogin J. State Department Cables Warned of Safety Issues at Wuhan Lab Studying Bat Corona Viruses. The Washington Post. April 14, 2020

24. Cumulative Confirmed COVID-19 Deaths by World Region Our World in Data. https://ourworldindata.org>grapher>comulative *cov ...*

Dr. Robert Marx

AFTERWARD

People call me "Give'm Hell Harry. Actually I never gave anyone hell. I just told them the truth and they thought it was hell." **Harry S Truman President of the United States 1945 – 1953**.

CLIMATE CHANGE: THE HOAX OF CO_2 REVEALED is the truth that the climate change obsessed thinks is hell. Their fixation on CO_2 ignores the real driving forces toward earth's temperature variations that nearly all nongovernmental independently funded climate experts agree. Their obsession has had serious consequences for the United States and our allies. It has stressed our economies to near recession levels. It has enriched and emboldened our adversaries such as China, Russia, Iran and Venezuela who ignore and thumb their noses at the notion of reducing fossil fuels and CO_2. Worse, it has allowed elected leaders to issue mandates that attempt to control what you drive, what you eat, what you can buy as well as introduced new unnecessary

Climate Change: The Hoax of CO_2 Revealed

taxes. While the CO_2 obsessed focus on CO_2 under the false banner of an "existential threat" they ignore and leave us all more vulnerable to real existential threats such as overpopulation, toxic pollution and pandemics. While real non-governmental scientists work to educate our citizens on the truth about Climate Change, we same citizens can combat the Climate Change obsessed lies in the voting booth.

The following list consists of 20 questions the climate change obsessed cannot answer. It is their hell.

1. What did CO_2 contribute to the ending of the last ice age and the melting of its glacier?
2. What did CO_2 contribute to the ending of the four previous ice ages?
3. Why was Al Gore's prediction of no ice left in the Arctic by 2015 wrong?
4. Why was James Hansen's computer models predicting no ice left in the arctic by 2017 wrong?
5. If reducing the CO_2 is the answer to global warming and there has been a 13% reduction in human CO_2 production, why is CO_2 still slowly increasing and the planet continuing to warm at a very slow rate?

Dr. Robert Marx

6. What really causes bleached coral?
7. Why did the author of the Hockey Stick graph write a retraction in the Nature Journal?
8. Why did the author of the Children's book "The Down to Earth Guide to Global Warming" reverse the data to falsely support global warming?
9. Why did NOAA misrepresent the degree of distance between the minimal and maximal elliptical paths of the earth's orbit around the sun?
10. What global warming reduction has been gained by the $500 million given to the UN by the Obama administration and the $1 billion given to the UN by the Biden administration for "Climate Change Reparations"?
11. Why does NOAA measure temperature readings next to heat exhaust fans as well as hot city buildings and pavement rather than in rural areas?
12. Why does NOAA measure CO_2 atop the Mauna Loa Volcano in Hawaii?
13. If the climate change obsessed believe CO_2 is the driving force of global warming, why is there no

Climate Change: The Hoax of CO_2 Revealed

governmental support for reforestation and the re-seeding of coral reefs?

14. How do the climate change obsessed explain the remains of a Viking farming village circa 1200 AD under the current Greenland Glacier?

15. How does a carbon tax or purchasing carbon credits reduce atmospheric CO_2?

16. If Methane from cows drives global warming, why is the current Biden administration not supporting red algae diet supplements known to reduce cow's methane emissions by 60% to 80%?

17. Is methane from any source really a problem when it represents only 0.00017% of the atmosphere?

18. What happens to the very flammable methane in the atmosphere during one of the 1.4 billion lightning strikes that occur globally each year?

19. Why do the 1609 climate scientists of the Non-InterGovernmental Panel on Climate Change (NIPCC) refute the data and very notion that fossil fuels and CO_2 are the cause of Global warming?

20. How do the climate change obsessed explain the fact that both ice cores and deep soil cores

Dr. Robert Marx

indicate that CO_2 concentrations have been more than ten times that of the present several times in the past?

Climate Change: The Hoax of CO_2 Revealed

Dr. Robert Marx

ACRONYMS

AR3 Third Assessment Report of the IPCC
AR5 Fifth Assessment Report of the IPCC 2013, 2014
AC Alternating Current
AD Anno Domini The Year of Our Lord
AIDS Acquired Immune Deficiency Syndrome
BCE Before Common Era
C Celsius
CaCo$_3$ Calcium Carbonate
CS$_2$ Carbon Disulfide
C$_6$H$_{12}$O$_6$ Glucose
CH$_4$ Methane
CO$_2$ Carbon Dioxide
COP Conference of the Parties
COVID-19 Coronavirus Disease 2019
CRU Climate Research Unit at the University of East Anglia
LA Niña Pacific Ocean Northern Oscillation
El Niño Pacific Ocean Southern Oscillation Pattern
EPA Environmental Protection Agency
EV Electric Vehicle

Climate Change: The Hoax of CO_2 Revealed

F Fahrenheit

FAO Food and Agricultural Organization of the United Nations

GCIG Global Climate Intelligence Group

GHCN Global Historical Climatology Network

GHG Green House Gases

GISS Goddard Institute for Space Studies

GRIP Greenland ICE Core Project

DC Direct Current

H1NI Hemagglutinin (a virus subtype)

Neuraminidase (a virus subtype)

HIV Human Immunodeficiency Virus

H^+ Hydrogen Ion (proton)

HCO_3^- Hydrogen Carbonate

H_2O Water (Dihydrogen Oxide)

H_2S Hydrogen Sulfide

IJC International Journal of Climatology

IPCC Intergovernmental Panel on Climate Change

Joules A unit of Energy

King Tide Annual Highest Tide

K2 Asteroid Asteroids that Course Close to Earth

LBS Pounds

LIA Little Ice Age

Mj/Kg Million Joules per Kilogram

MWP Medieval Warm Period

N₂ Nitrogen

NASA National Aeronautics and Space Administration

NIPCC Nongovernmental International Panel on Climate Change

NOAA National Oceanic and Atmospheric Administration

O₃ Ozone

O₂ Oxygen

OMFP Oral and Maxillofacial Pathologist

OMFS Oral and Maxillofacial Surgeon

Pg Pentagrams 10^{15}

pH Potential of Hydrogen. A measure of Acidity or Alkalinity

PITCH A human Gene that produces the PITCH-1 Protein for Normal Development

PPM Parts per Million

RATPAC Radiosonde Atmospheric Temperature Products for Assessing Climate

RSV Respiratory Syncytial Virus

SARS Severe Acute Respiratory Syndrome

Climate Change: The Hoax of CO_2 Revealed

SL Sea Level

SST Sea Surface Temperature

TSI Total Solar Irradiance

UN United Nations

UNFCC United Nations Framework Convention on Climate Change

USECC United States Envoy For Climate Change

W/m² Watts per Meters Square

WCD World Climate Intelligence Group

WHO World Health Organization

Y2K Year Two Thousand

www.ingramcontent.com/pod-product-compliance
Lightning Source LLC
Chambersburg PA
CBHW052247220526
45471CB00001B/234